Amritesh Kumar

Micromechanical study on the deformation behaviour of directionally solidified NiAl-Cr eutectic composites

Schriftenreihe
des Instituts für Angewandte Materialien
Band 69

Karlsruher Institut für Technologie (KIT)
Institut für Angewandte Materialien (IAM)

Eine Übersicht aller bisher in dieser Schriftenreihe erschienenen Bände finden Sie am Ende des Buches.

Micromechanical study on the deformation behaviour of directionally solidified NiAl-Cr eutectic composites

by
Amritesh Kumar

KIT Scientific Publishing

Dissertation, Karlsruher Institut für Technologie
KIT-Fakultät für Maschinenbau

Tag der mündlichen Prüfung: 02. März 2017
Referenten: Prof. Dr. rer. nat. Oliver Kraft, Prof. Dr. Anton Möslang

Impressum

KIT Scientific Publishing

Karlsruher Institut für Technologie (KIT)
KIT Scientific Publishing
Straße am Forum 2
D-76131 Karlsruhe

KIT Scientific Publishing is a registered trademark
of Karlsruhe Institute of Technology.
Reprint using the book cover is not allowed.

www.ksp.kit.edu

Print on Demand 2017 – Gedruckt auf FSC-zertifiziertem Papier

ISSN 2192-9963
ISBN 978-3-7315-0694-2
DOI 10.5445/KSP/1000071898

Micromechanical study on the deformation behaviour of directionally solidified NiAl-Cr eutectic composites

Zur Erlangung des akademischen Grades
Doktor der Ingenieurwissenschaften
der Fakultät für Maschinenbau
Karlsruher Institute für technologie (KIT)

genehmigte
Dissertation
von

Dipl. Ing. **Amritesh Kumar**
aus Patna, Indien

Tag der mündlichen prufung:	02.03.2017
Hauptreferent:	Prof. Dr. rer. nat. Oliver Kraft
Korreferent:	Prof. Dr. Anton Möslang

Abstract

Nickel aluminide (NiAl) is considered to be a promising candidate as next generation material for high temperature structural application owing to a range of attractive properties, but suffers from low ductility and fracture toughness at room temperature and poor creep resistance and strength at elevated temperatures. One of the most promising ways to improve its properties is by alloying binary NiAl with refractory metals such as Mo, Cr, V, Hf, Re etc. to produce pseudo binary eutectic alloys and further processing of these alloys by directional solidification which results in highly aligned in-situ composites having superior properties compared to binary NiAl. The alloying element used and the processing conditions chosen for the directional solidification has a great impact on the final microstructure of the alloy, which in-turn affects the mechanical properties of the alloy produced. Although a great amount of research has been done with different alloying elements and processing conditions, optimum parameter and conditions to achieve the best properties has not been established yet, neither is the deformation mechanisms in these alloys fully understood.

The main aim of this thesis is to characterize the mechanical behaviour of directionally solidified NiAl-Cr eutectic at varying length scales in order to develop a mechanistic understanding of role of individual phases and interfaces on the deformation process in these alloys. NiAl-Cr eutectic alloy directionally solidified at 20, 50 and 80 mm/h were studied in this study. The individual phases and the interfaces were tested by different micromechanical tests to study their influence on the deformation processes.

Microstructure examination shows that the fiber diameter and fiber spacing decreases with the increasing solidification speed. Nanoindentation study on the overall alloy reveals that the hardness and young's modulus of these alloys do not depend on the solidification speed. In-situ SEM tensile tests on isolated individual Cr fibers shows that they deform plastically before failure and show high strength values in the range of 1-3 GPa. Compression tests on three different types of micro-pillars produced by focused ion beam milling reveals that while with increasing solidification speed and thus decreasing pillar diameter, the strength of each type of pillar increases, the matrix only (M.O.) pillars are stronger than the single fiber (S.F.) composite pillars having the interface for similar dimensions of these micro-pillars. TEM examination of the undeformed specimen shows that both the NiAl matrix and Cr fibers have very low dislocation density. The interface between the fiber and matrix is surrounded by the network of <100> type dislocations due to the small lattice mismatch between NiAl and Cr. The deformed S.F. pillars show heavy dislocation activity both in the fiber and the matrix. Most of these dislocations were found to have a <100> Burgers vector. A rule of mixture analysis based on the strength of Cr fiber and matrix only (M.O.) pillars predicts higher strength values for S.F. pillars than that observed in the experiments. Cross cuts across the deformed S.F. pillars shows that the two phases co-deform upon straining without any delamination across the interface suggesting a strong interface. All the observations from the different tests performed strongly suggest that the interface between the fiber and matrix is dominating the deformation of these directionally solidified NiAl-Cr eutectic alloys by providing additional mobile dislocation generation sites at the interface. This would also explain the independence of hardness with respect to the solidification speed where the increase in strength of Cr fibers and NiAl matrix with decreasing size and spacing with increasing solidification speed, is counteracted by the increase in the interface area per unit volume leading to lower strength and higher ductility.

Kurzzusammenfassung

Nickelaluminide (NiAl) werden als vielversprechender Kandidat für Hochtemperatur-Strukturmaterialien der nächsten Generation aufgrund einer Reihe attraktiver Eigenschaften betrachtet. Allerdings besitzen diese Materialien eine geringe Duktilität und Bruchzähigkeit bei Raumtemperatur, sowie schlechte Kriecheigenschaften und eine geringe Festigkeit bei höheren Temperaturen. Eine vielversprechende Strategie zur Verbesserung dieser Eigenschaften stellt das Zulegieren von binärem NiAl mit Refraktärmetallen, wie z. Bsp. Mo, Cr, V, Hf, Re, usw. dar. Diese pseudo-binären Legierungen werden durch gerichtetes Erstarren produziert, was zu hochausgerichteten in-situ Werkstoffverbunden mit verbesserten Eigenschaften im Vergleich zu binärem NiAl führt. Die Wahl der Legierungselemente und der Prozessbedingungen hat einen großen Einfluss auf das Gefüge der Legierung, was wiederum die mechanischen Eigenschaften beeinflusst. Obwohl der Einfluss verschiedener Legierungselemente, als auch unterschiedlicher Prozessbedingungen schon untersucht wurden, konnten bisher weder die optimale Parameter und Bedingungen bestimmt werden, noch sind die Deformations-mechanismen in diesen Legierungen vollständig verstanden.

Das Hauptziel dieser Arbeit ist die Untersuchung des mechanischen Verhaltens eutektisch richtungserstarrter NiAl-Cr Legierungen auf unterschiedlichen Längenskalen um ein systematisches Verständnis der Rolle der einzelnen Phasen und der Grenzflächen auf die Verformungsvorgänge in diesen Legierungen zu erhalten. Eutektische NiAl-Cr Legierungen, die bei Erstarrungsgeschwindigkeiten von 20, 50 und 80 mm/h gerichtet hergestellt wurden, wurden in dieser Arbeit untersucht. Die einzelnen Phasen und Grenzflächen wurden mittels verschiedener mikromechanischer Versuche getestet, um ihren Einfluss auf die Verformungsvorgänge zu bestimmen.

Gefügeuntersuchungen haben gezeigt, dass der Faserdurchmesser und abstand mit höheren Erstarrungsgeschwindigkeiten sinkt. Nanoindentationsergebnisse an den Proben haben ergeben, dass sowohl die Härte als auch der Elastizitätsmodul dieser Legierungen nicht von der Erstarrungsgeschwindigkeit abhängen. In-situ Zugversuche in einem REM an einzelnen Cr-Fasern haben gezeigt, dass die Fasern sich plastisch vor dem Versagen verformen und Festigkeiten im Bereich von 1-3 GPa erreichen. Druckversuche an drei unterschiedlichen Arten von Mikrosäulen hergestellt mittels FIB zeigen, dass mit wachsender Erstarrungsgeschwindigkeit und dadurch kleiner werdenden Säulendurchmessern, die Festigkeit des jeweiligen Säulentyps ansteigt. Säulen, die aus der Matrix geschnitten wurden, sind fester als jene, die eine einzelne Faser und somit mehr Grenzflächen bei gleichbleibender Säulengröße beinhalten. TEM Untersuchungen an den unverformten Proben haben gezeigt, dass sowohl die NiAl Matrix, als auch die Cr-Fasern eine sehr niedrige Versetzungsdichte besitzen. Die Grenzflächen zwischen Faser und Matrix sind von einem Netzwerk von <100> -Versetzungen aufgrund der geringen Gitterfehlpassung von NiAl und Cr, umgeben. Die formierten Säulen mit einer einzelnen Faser zeigen hohe Versetzungsaktivitäten sowohl in der Faser, als auch in der Matrix. Eine Analyse anhand einer Mischungsregel basierend auf den Festigkeiten der Cr-Fasern und der reinen Matrix sagt höhere Festigkeiten für die Säulen mit einer einzelnen Faser voraus, als in den Experimenten beobachtet wurde. Querschnitte durch die deformierten Säulen mit einer Faser haben gezeigt, dass beide Phasen unter der Belastung ohne Delamination kodeformieren, was auf eine starke Grenzfläche hinweist. Alle Beobachtungen der unterschiedlichen Versuche deuten darauf hin, dass die Grenzfläche zwischen Faser und Matrix das Verformungsverhalten der eutektisch richtungserstarrten NiAl-Cr Legierungen dominiert, indem sie zusätzliche, bewegliche Versetzungen. Dies würde ebenso die unveränderte Härte bei verschiedenen Erstarrungs-geschwindigkeiten erklären, da dem Festigkeitszuwachs der Cr-Fasern und der NiAl-Matrix mit abnehmender Größe und Abständen der Zugewinn an Grenzfläche pro Volumeneinheit entgegenwirkt, was schließlich zu niedrigeren Festigkeiten und höherer Duktilität führt.

Acknowledgments

This doctoral thesis has originated from my work at the Institute for Applied Materials (IAM-WBM) at Karlsruhe Institute of Technology from January 2013 till December 2016 under the framework of Helmholtz graduate school 'Integrated materials development for novel high temperature alloys (IMD)'. I would like to thank everyone who has supported me in successfully completing my Ph.D. thesis.

First, I would like to thank my thesis advisor Prof. Dr. Oliver Kraft, who as the institute leader and spokesperson for IMD gave me the opportunity to obtain my doctorate at the institute. His constant input and ideas and our discussions provided important contribution towards this work. This thesis would not have been possible without his constant support and encouragement.

I would like to express my gratitude for Dr. Ruth Schwaiger for all the support and encouragement during my time at the institute. Apart from our many discussions and her helpful advices, I have learned a lot from her professionalism and enthusiasm towards scientific work.

My sincere thanks goes to Prof. Martin Heilmaier, Dr. Heinz Riesch-Oppermann and all the colleagues of our graduate school 'IMD' for creating the environment for collaborative work along with scientific advice and discussions.

I would also like to thank Ms. Daniela Exner and Mr. Ewald Ernst for their help in samples preparation and experimental work.

Working at IAM-WBM was a wonderful personal and cultural experience for me and for this I would like to thank all my colleagues at the institute for creating a wonderful work environment.

I would like to express my deepest gratitude to my parents, my sister and brother-in-law, and my sweet nieces who has supported me at every step of my life and has been a source of strength and happiness for me.

Finally, I would like to thank my long-time friend and now my wife Nilesha Mishra for all her love and support.

List of figures

List of tables

Table of Contents

1 Introduction

1.1 High temperature structural application

High temperature application is critical for various industries such as materials processing, production of metals and non-metals, chemical engineering, transportation, energy conversion and power generations. The operating temperature in many of these applications are above 600 °C where most steels cannot be used anymore and can go as high as 4000 °C for rocket nozzles. Typical operating temperatures of few of these applications are shown in Figure 1.1 as a plot with respect to temperature. In particular for engines, the cost of fuel is a big concern and there is a constant effort to increase their efficiency. Higher efficiency leads to same amount of work done with less fuel usage or more work done with same amount of fuel. The efficiency of these applications increases with increasing the operating temperatures as shown in Figure 1.2 for a gas turbine engine, for example. As the temperature increases, the green curve shows that the ideal power generated by the gas turbine engine increases with increasing temperature (Perepezko, 2009). This increased efficiency leads to lower fuel consumption and lower exhaust emission, thus giving benefits economically as well as environmentally. But as also shown in Figure 1.2, the development trend of the new materials and alloys for these high temperature engine applications, which shows that there exist inefficiency losses as compared to the ideal curve. These losses tend to increase with increasing temperature if the same development trend is maintained. These losses arise mainly because the alloy used in these applications cannot withstand the high temperatures and pressure unless they are aided by a cooling system. Therefore, we need to develop new materials and alloys to close the gap arising from inefficiency losses and to further increase the efficiency of the high temperature applications.

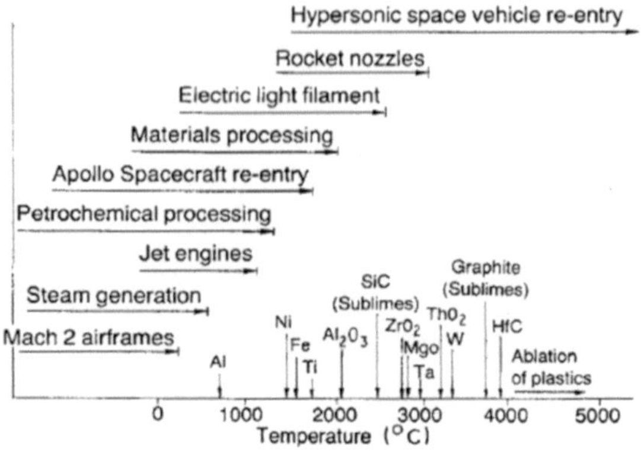

Figure 1.1: Typical operating temperatures for few high temperature applications (Meetham, 1991)

High temperature operation is associated with various challenges (Anton, Shah, Duhl, & Giamei, 1989; Meetham, 1991), and the material selected for these application typically fulfil certain criteria. The operating temperature for most materials should not exceed 0.5 Tm, the melting temperature, thus the selected material must have a high melting point. The selected material must also have very good strength at higher temperature to withstand the stresses generated in the components. In general, some ductility is also required to avoid any catastrophic failure during processing and application of these alloys. Oxidation and corrosion are a major concern at higher temperature and it is desired to have a material with high oxidation and corrosion resistance to retain its properties in oxidizing atmosphere. Good oxidation resistance is also required to minimize the need of protective coatings which may have a negative impact on the efficiency of the application. The materials should have good structural stability and very good creep and fatigue properties as these materials are required to serve at high temperature and cyclic stresses for a considerable long lifetime. Processing these alloys in the desired form or shape also induces very high stresses and it is generally done at room temperatures. Therefore, it is necessary to have sufficient toughness and ductility at room temperature to avoid generation of micro crack during processing, which would otherwise lead to poor performance. Furthermore, since many of these applications involve the dynamic use of components, the specific strength and thus the density of the material becomes a very important criterion. And lastly, the cost of raw materials and processing of these alloys must be reasonable to ensure mass production of these components.

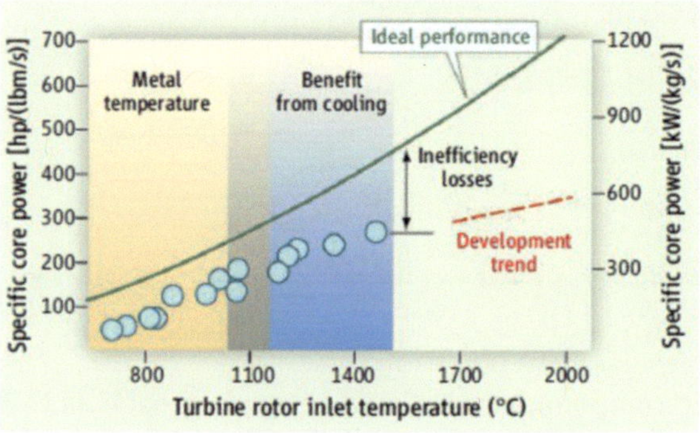

Figure 1.2: Variation of specific core power produced in a turbine engine, showing ideal performance along with development trend of turbine engines (Perepezko, 2009)

Development of new materials which can push the limits of operation of these applications while satisfying the above mentioned critical and often contradicting criteria is a very important challenge for material scientists today. Materials development for these high temperature application is a continuous process. A number of high temperature alloys have been developed over the past decades with different properties to suit different applications (Meetham, 1991)(Birol, 2011; Pollock & Tin, 2006). The most notable are aluminium alloys, steels, iron based superalloys, cobalt based superalloys, nickel based superalloys and ceramics. A comparison of various alloys developed over the years with respect to the temperature capabilities of these alloys is represented in Fig.1.3.

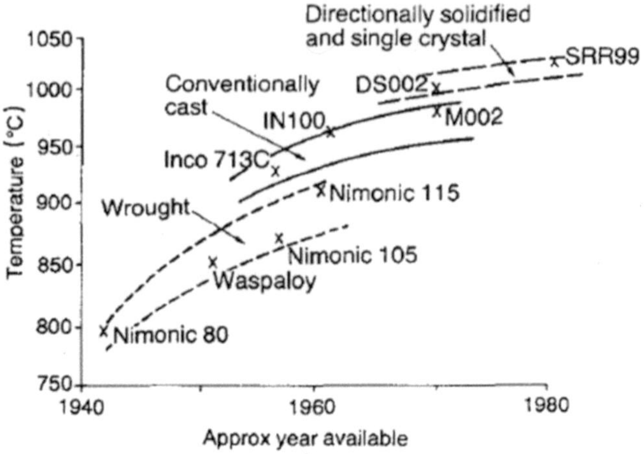

Figure 1.3: Development of alloys over the years for high temperature applications *(Reed, 2006)*

Intermetallic compounds containing aluminium such as NiAl, TiAl, FeAl and others are among the promising candidates to replace the currently used materials in these high temperature structural applications in order to increase the temperature capabilities of the application. Aluminides are promising in terms of being high strength as well as low density alloys. NiAl is one of the most promising of these aluminides having a range of attractive properties, high strength, low density, high thermal conductivity, good oxidation resistance to name a few. NiAl also suffers from the typically associated drawbacks, having poor ductility and fracture toughness at room temperature and low creep resistance at high temperature.

A lot of research has been directed in the past towards elimination of these drawbacks of NiAl (W. Chen, Hines, & Wang, 2004; Darolia, 1991; Deevi & Sikka, 1996; Daniel B. Miracle & Darolia, 2000; Wang, Qian, Zhang, & Wang, 2011), producing single crystals, alloying with refractory metals and processing by novel techniques and more. Alloying NiAl with refractory metals such as Cr, Mo, V or Re to produce pseudo-binary eutectics and processing of these eutectics by directional solidification result in highly aligned fiber-matrix composite microstructures (Bei, Pharr, & George, 2004; Georg Frommeyer & Rablbauer, 2008; David R Johnson, 1994). They have been shown to improve the fracture toughness and creep resistance compared to binary NiAl. Based on the alloying element and processing conditions chosen for directional solidification in these eutectics, a variety of final microstructures can be produced. However, a detailed materials development with respect to alloying additions and processing condition to get the best combination of properties has not been established yet. This is due to the fact that the role of alloying addition and the microstructures-property relationship has not been fully understood yet.

1.2 Outline of the thesis

The thesis is organized in the following way. The most important literature, relevant for this project is reviewed in Chapter 2, including the knowledge base available for NiAl, directionally solidified eutectics, size effects in materials at small scale as well the different toughening approaches and their mechanisms. In Chapter 3, details of the various test methods used in this project are explained including the description of the experimental set-ups, sample preparation and testing and analysis procedures. In Chapter 4, the results from the different mechanical tests performed are presented. Key findings and observation from these results are cross-linked and discussed in order to develop a comprehensive understanding of the deformation mechanism in the directionally solidified eutectics. And finally, in Chapter 5, the thesis is summarized and a short outlook is given.

2 State of the art

2.1 Binary NiAl and its properties

The two main aluminides in the Ni-Al system which are of significance are Ni_3Al and binary NiAl (Fig. 2.1). Ni_3Al has received a great amount of attention in the past years for structural materials as a strengthening phase in most of the superalloys developed in recent past for high temperature structural application. NiAl had emerged as a promising candidate in replacing Ni-based and other superalloys about 30 years ago. The main reason for this interest its advantages over superalloys in terms of density, melting temperature, oxidation and corrosion behaviour among others (Darolia, 1991; D. B. Miracle, 1993).

The physical and mechanical properties of NiAl have been studied in detail over the years and a comprehensive review of the physical and mechanical properties of NiAl is presented by (D. B. Miracle, 1993), (Noebe, Bowman, & Nathal, 1993) (Dey, 2003).

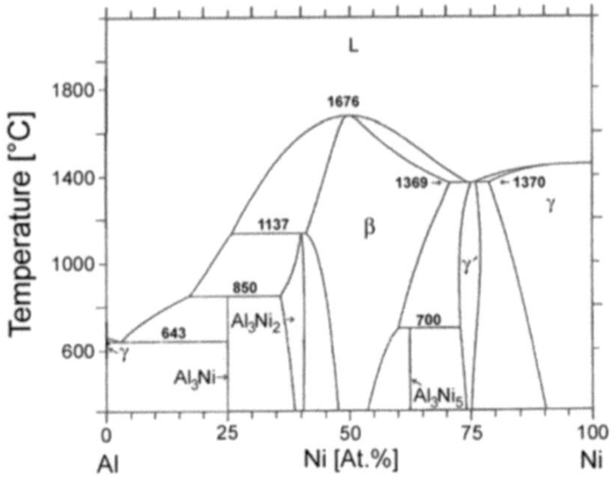

Figure 2.1: Phase diagram of binary NiAl *(Ansara, Dupin, Lukas, & Sundman, 1997)*

2.1.1 Crystal structure and physical properties

Binary NiAl exists over a range of compositions in the Al-Ni phase diagram from 45 to about 60 at.-% Ni (Fig. 2.1). Stoichiometric NiAl exists at Ni- 50 Al (at. %). The crystal structure of NiAl is primitive cubic CsCl structure. The strukturbericht designation for this structure is B2 (Villars, Calvert, & Pearson, 1985), which is described as two interpenetrating cubic lattices where Al atoms occupy one sublattice and Ni atoms occupy the other as shown in Fig. 2.2. Stoichiometric NiAl congruently melts at 1911 K. The lattice parameter of NiAl is 0.2887nm (A. Taylor & Doyle, 1972). Since NiAl exists over a range of compositions, the lattice parameter varies strongly with respect to composition. The maximum of lattice parameter occurs at stoichiometric composition and goes down on either side.

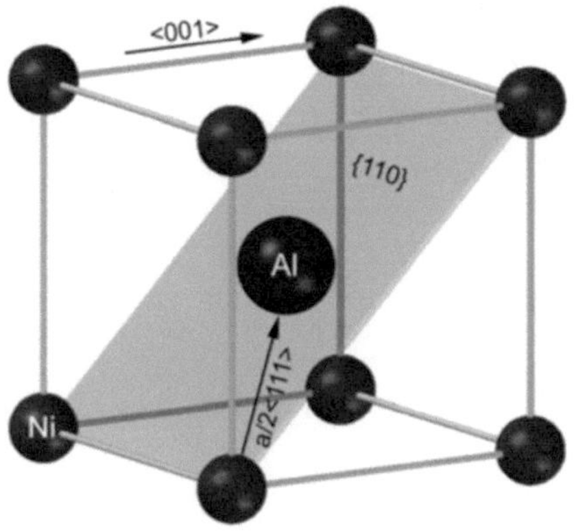

Figure 2.2: B2 ordered unit cell of NiAl (Georg Frommeyer & Rablbauer, 2008)

The density of NiAl decreases linearly with increasing Al concentration, though a change in slope of the density variation occurs at stoichiometric composition which has a density of 5.90 g/ cm^3 (A. Taylor & Doyle, 1972). Stoichiometric NiAl has excellent phase stability when compared to other intermetallic compounds with B2 structure, because of having one of the largest heats of formation (Kubaschewski, 1958). It has been observed that significant ordering exists in NiAl upto the melting temperature. It has also been observed that NiAl is also very stable against radiation damage (Mori, Fujita, Tendo, & Fujita, 1984), which makes it a promising candidate for nuclear applications.

The elastic properties of single crystal NiAl have been found to be very similar to the polycrystalline values. The Young's modulus values of NiAl is observed to be very sensitive to the temperature and heat treatments, while composition has almost no effect on the values. The Young's modulus of NiAl has been reported to be around 190 GPa (Rusović & Warlimont, 1977)(Rusović & Warlimont, 1979). The Young's modulus of NiAl decreases linearly with the increase in the temperature. Other physical properties of NiAl also depend strongly on the temperature. The list of important physical properties is tabulated in the table 2.1.

Table 2.1: Important Physical properties of NiAl *(Dey, 2003)*

Properties	NiAl
Electrical resistivity (10^{-8} Ωm)	8-10
Thermal conductivity (W/m.K)	76
Thermal expansion coefficient ($10^{-6}/K^{-1}$)	13.2
Lattice parameter (nm)	0.2887
Young's modulus (GPa)	190
Specific heat (J/g.K)	0.64
Melting point (K)	1911

2.1.2 Mechanical properties

Operative slip system

Several studies have been performed to determine the operative slip systems in near stoichiometric single crystalline NiAl as well as polycrystalline NiAl both theoretically and experimentally. Some important theoretical studies with respect to slip in NiAl are discussed ahead. Rachinger and Cottrell (Rachinger & Cottrell, 1956) made the theoretical prediction on the basis of ordering energy and predicted <001>{110} slip system in NiAl. Lautenschlager et al. (Lautenschlager, Hughes, & Brittain, 1967) analysed slip modes in terms of a hard sphere model suggesting that <001>{110} and <001>{100} should be the primary slip system in NiAl.

Ball and Smallman (Ball & Smallman, 1966b) used anisotropic elasticity theory and predicted the operative slip system to be <100>{011}. Yoo et al. (Yoo, Takasugi, Hanada, & Izumi, 1990) studied NiAl system using energetic and kinetic aspect of dislocation motion and have concluded that the <001> slip should occur in NiAl. To conclude, all theoretical study in case of NiAl, regardless of technique, agree that NiAl should have a<100> slip vector and {011} slip plane with the possibility of slip occurring on {001} planes.

Experimental observations suggest that the NiAl single crystals slip behaviour strongly depends on the orientation with soft orientation and hard orientations. All the non- [001] orientated NiAl single crystals are termed as *soft* orientation as compared to *hard* [001] oriented NiAl single crystals. A number of experimental studies and techniques have been employed to identify the slip system in both hard and soft oriented NiAl single crystals.

In soft orientations, for example, Pascoe and Newey (Pascoe & Newwy, 1968) identified a <001> slip vector by slip trace analysis, Ball and Smallman (Ball & Smallman, 1966a)(Ball & Smallman, 1966b) observed <001> {110} slip by TEM. Cross slip and pencil glide on {110} plane was also observed. In almost all the studies, the only observed slip systems have been <001> {110} and <001> {100} and are independent of the employed temperature of deformation.

In case of hard-[001] orientations, the resolved shear stress reaches almost zero and stress required to plastically deform the specimen is several times higher than for the soft orientations at low and intermediate temperatures (R. R. Bowman & Noebe, 1992). At low temperatures, the observed slip direction is <111> on either {123}, {112} or {-110}. From room temperature till ~600 K both kinking and deformation by <111> on {-1 -1 2} and {-110} planes have been observed. At higher temperatures, above 600 K deformation is observed to be the result of either <001> type dislocation, <011> type dislocation or a combination of climb and glide on both (Noebe et al., 1993). Studies on the polycrystalline NiAl with respect to the slip systems (R. Bowman, Noebe, Raj, & Locci, 1992) conclude that the deformation of polycrystalline NiAl occurs by <001> dislocation operation.

Ductility

The room temperature ductility of single crystals is anisotropic and depends on the orientation. The ductility of single crystalline NiAl also depend on the stoichiometry, interstitial content, impurities. 'Hard' single crystal NiAl fail after only elastic elongation, while soft single crystals show plastic elongations up to about 2% (Lahrman, Field, & Darolia, 1990). Inadequate dislocation sources, low dislocation mobility and low fracture stress are some of the factors considered to be responsible for the low ductility of the NiAl single crystals. The DBTT of hard single crystal is around 673 K while the soft single crystal become ductile at around 473 K (Lahrman et al., 1990).

The room temperature ductility of polycrystalline NiAl is generally very limited and is dependent on the stoichiometry, grain size, texture, impurities etc. As we have already discussed about the operative slip systems in binary NiAl, the <001> slip direction provides for only three independent slip systems. This limitation has been argued as the main reason of the limited ductility as five independent slip systems are generally considered the minimum requirements for reasonable ductility (Ball & Smallman, 1966b). The ductile to brittle transition temperature DBTT of polycrystalline NiAl is around 700 K, above which polycrystalline NiAl becomes very ductile and plastic elongations of around 40% have been observed at 873 K (Hahn & Vedula, 1989). It is believed that both climb and glide are responsible for the increased ductility as additional slip systems have not been observed to operate at higher temperature.

Strength

The yield strength of stoichiometric polycrystalline NiAl and 'soft' NiAl single crystal is very low compared to existing Ni-based superalloys used today and therefore cannot be considered to be used in high-temperature structural applications in an un-alloyed state. The yield strength of NiAl single crystal in soft and hard orientation and NiAl polycrystals are shown in Fig. 2.3 along with strengths of existing Ni-based superalloys. NiAl single crystals in 'hard' orientation on the other hand shows reasonable strength values (Darolia, 1991)(Hack, Brzeski, & Darolia, 1995). The strengths of 'hard' single crystal NiAl, although being superior to 'soft' NiAl, is strongly dependent on the stoichiometry and temperature. The strengths drops rapidly with increasing temperature (Pascoe & Newwy, 1968). Deviation from stoichiometric composition leads to strong increment in the strength levels. But this effect also diminishes with the increase in temperatures and at around 1400K, the strength levels of stoichiometric NiAl is similar to non-stoichiometric NiAl (Pascoe & Newwy, 1968).

Fracture properties

The single crystal fracture toughness of NiAl depends on the crystallographic directions, specimen configuration and geometry of the notch. Specimens with straight notch tested in <110> orientation have K_{IC} values of 4-5 MPa-$m^{1/2}$ (Reuss & Vehoff, 1990) while, for <100> orientation, it is around 8 MPa-$m^{1/2}$ (Chang, Darolia, & Lipsitt, 1992). The fracture toughness of the single crystals also increases with increasing temperatures to about 15 MPa-$m^{1/2}$ near 500 K for <110> specimens and to about 30 MPa-$m^{1/2}$ for <100> specimens.

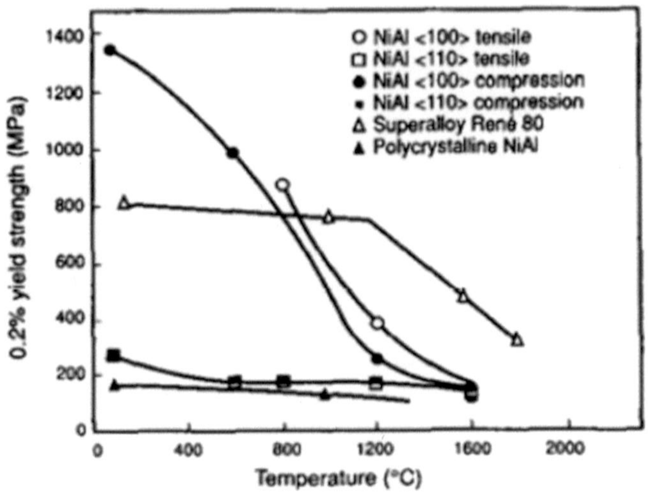

Figure 2.3: Yield strength of single crystalline and polycrystalline NiAl in 'hard' and 'soft' orientation v/s temperature (Darolia, 1991)

Fracture of polycrystalline NiAl is predominantly intergranular at RT (Hahn & Vedula, 1989), which changes to transgranular cleavage at around 600 K. Above 900 K, the fracture occurs by ductile fracture. The fracture toughness (K_{IC}) of polycrystalline binary NiAl is ~ 4-6 MPa-$m^{1/2}$ at room temperature (Kumar, Mannan, & Viswanadham, 1992)(H. E. Cline, Walter, Lifshin, & Russell, 1971). The fracture toughness increases with increasing temperature as a result of increased crack tip plasticity at elevated temperatures. The processing of NiAl also affect the fracture behaviour and while zone-refined NiAl shows K_{IC} values of around 50 MPa-$m^{1/2}$ at 673 K, the values for as-cast NiAl values increase to only 10 MPa-$m^{1/2}$ (Reuss & Vehoff, 1990).

To summarize, although having excellent physical properties, the mechanical properties of NiAl, especially with respect to ductility and fracture toughness at room temperatures, need improvement for replacing Ni-based superalloys. The poor room temperature ductility and fracture toughness are mainly attributed to the lack of sufficient number of independent number of slip systems in NiAl. There has been considerable amount of research in the past focusing on improving the properties of NiAl, producing single crystals to eliminate grain boundaries, alloying with other elements (metals, non-metals and intermetallics), as well as different processing methods to produce high purity alloys (R. Bowman et al., 1992; Darolia, 1991; Georg Frommeyer & Rablbauer, 2008; Daniel B. Miracle & Darolia, 2000; Noebe et al., 1993). One of the most promising of these approaches is to alloy binary NiAl with refractory metals producing eutectic alloys, and processing of these eutectic alloys by directional solidification, which is discussed in the following.

2.2 NiAl-X directionally solidified eutectics

Alloying with refractory metals such as Cr, Mo, Re, V to produce eutectic alloys have been shown to improve the toughness and creep strength from binary NiAl values (W. Chen et al., 2004; C. T. Liu & Horton, 1995; Pank, Koss, & Nathal, 1990; Rablbauer, Fischer, & Frommeyer, 2004).

For further improvements in the properties, processing of these eutectic alloys by directional solidification results in in-situ composite microstructures where the phases are aligned in the growth directions (Bei & George, 2005; Bei et al., 2004; Haenschke et al., 2010; D. R. Johnson, Chen, Oliver, Noebe, & Whittenberger, 1995a; Misra, Wu, Kush, & Gibala, 1998; Whittenberger, Raj, Locci, & Salem, 1999; Yang, 1997a). These *in-situ* composites, in which the phases are formed simultaneously during the composite fabrication, are differentiated from *ex-situ* composites, in which a second phase is produced separately and inserted into the matrix externally after the solidification of the matrix phase.

A number of key factors must be considered in the directional solidification of the alloys, such as alloy composition, growth velocity and undercooling at the solidification front. Depending on the set of parameters chosen for directional solidification, different composite microstructures are obtained which can either have a rod-like or lamellar morphology. Jackson and Hunt (Jackson & Hunt, 1966) have discussed in detail, the role of these factors on the growth structures of these eutectics. The details of directional solidification process is not discussed in this thesis and readers are referred to book "Fundamentals of solidification" (Kurz & Fisher, 1986) for further information.

This study has been focused on directionally solidified NiAl-Cr eutectic alloy produced at different solidification velocities. Cr was selected as the alloying element as it is one of the lightest elements among the various refractory metals identified in order to improve the mechanical properties of binary NiAl. A lot of research has been conducted on the effect of solidification parameters of ternary NiAl-Cr eutectic. Cline and Walter (Walter & Cline, 1973) have investigated the effects of alloying additions and processing conditions and relative stabilities of these microstructures. They have shown that transition from rod-like to plate-like morphology in NiAl-Cr eutectic starts with addition of 1.3% V and, at 5%, it completely changes to lamellar structure, while in case of tungsten addition, the transition starts at 0.7%. Similarly, replacing 0.6% Cr from NiAl-Cr eutectic by same amount of Mo causes the formation of faceted Cr(Mo) rods and plates instead of non-faceted rods. Walter and Cline (Walter & Cline, 1970) have also investigated the effect of the solidification rate on the microstructure of these eutectics. They showed that for the NiAl-Cr eutectic the spacing between the fibers decreases with solidification speed as $V^{-1/2}$, where V is the solidification speed. Also increasing the solidification speed above 1 inch/hr results in the formation of a cellular structure and the number of cells increases with the solidification speed. The alloying additions and processing conditions, by affecting the resulting microstructure of the eutectics, affects the properties of the alloy. Several studies have been performed on the mechanical properties of different ternary and quaternary alloys of NiAl-Cr and NiAl-Cr(X), X being Mo, Hf, Re, V, in relation to the solidification speed, final microstructure, and faults in the structure. Walter and Cline (Walter & Cline, 1970) examined the effect of alloying addition and solidification speed on the high temperature strength of NiAl-Cr alloys and concluded that the tensile strength at 600 °C increases greatly with decreasing the inter-fiber spacing. At higher temperatures of 800 °C and 1000 °C the strength first increases and then decreases with decreasing fiber spacing, while at 1200 °C the strength decreases slightly with inter-fiber spacing.

Johnson and co-workers (D. R. Johnson, Chen, Oliver, Noebe, & Whittenberger, 1995b) studied the mechanical properties of NiAl-Cr and NiAl-(Cr, Mo) eutectic alloys with different growth velocity. They have observed a fiber-matrix composite structure for NiAl-Cr and NiAl-(Cr, Mo) up to 0.6 % of Mo. Adding more than 0.6 % of Mo transforms the microstructure to a lamellar microstructure. The NiAl-Cr and NiAl-(Cr-Mo) samples both showed improved strength and creep properties compared to binary NiAl, although when compared to Ni-based superalloys, these alloys were found to be much weaker. Similarly, the fracture toughness of these eutectics were found to be significantly higher than binary NiAl, 20 MPa-$m^{1/2}$ for DS eutectics as compared to 6 MPa-$m^{1/2}$ for the polycrystalline NiAl or 11 MPa-$m^{1/2}$ for single crystal NiAl. The fracture morphology and different mechanisms responsible for the increased fracture toughness in these DS eutectic alloys are discussed in detail in their publication.

Whittenberger and co-workers (Whittenberger et al., 1999) have discussed the room temperature fracture properties of NiAl-31Cr-3Mo, which is directionally solidified at a range of solidification velocities from 7.6 mm/h to 508 mm/h. The samples had a lamellar microstructure, although it was noted that except for a solidification speed of 12.7 mm/h, all the samples showed cellular microstructure enclosed by intercellular regions. They found that the RT fracture toughness values fall in two groups; at lower solidification speeds, the fracture toughness values lie around 12 MPa-$m^{/12}$ while increasing the solidification speed further resulted in higher fracture toughness values of around 17 MPa-$m^{1/2}$. They have also found that the NiAl-31Cr-3Mo alloys produced with growth velocities higher than 25 mm/h had similar mechanical properties.

Suffice to say, the best suited alloying elements and corresponding solidification speeds to achieve the optimum mechanical properties in NiAl based directionally solidified eutectic alloys is yet to reach a consensus. As explained above, Walter and Cline and Johnson and co-workers among others, have reported significant effect of alloy composition and processing conditions on the mechanical properties of these DS eutectics, while Whittenberger and co-workers report that above certain growth speeds the mechanical properties are nominally similar. Therefore, it becomes very important to systematically study these directionally solidified eutectics and efforts must be made to develop our understanding of deformation mechanisms of these eutectics and the role of the individual phases and interfaces.

Several studies have been focused towards developing a better understanding of the micromechanics of deformation of these eutectics. Cline et al. (E. Cline, Walter, Koch, & Osika, 1971) have studied the dislocation structures at the interface of NiAl and Cr fibers in as prepared directionally solidified NiAl-Cr eutectic. They have observed that the interface between the NiAl matrix and the Cr fiber consists of a network of interface dislocation which surrounds the Cr fiber along the curved interface. These dislocation network is believed to be formed to reduce the internal stresses caused during the solidification process arising due to small difference in the lattice parameters of the two phases. These dislocations were found to be mostly of a<100> type edge dislocation, although some parts of these interface dislocation network may contain dislocation of other Burgers vectors. They also estimated the lattice mismatch between the two phases based on the measurement of the dislocation spacing in the network to be around 0.35 %. It must be stated that this value is based on the equilibrium lattice parameters and the actual value of lattice mismatch may differ from the equilibrium values.

Other studies focusing on studying the dislocation structures and their contribution to the deformation and fracture behaviour have been performed. Misra and Gibala (Misra & Gibala, 2000) for example have studied the dislocation evolution in the deformation of NiAl-34(at. %) Cr and NiAl-

9(at. %) Mo alloys. They examined the dislocation structures in the as prepared, low deformed and high deformed alloys. The as-prepared NiAl-Cr eutectic alloy, show a cube on cube orientation relationship between NiAl matrix and Cr fibers, with growth direction <001> in both phases. The interface is reported to be semi-coherent with a network of interface dislocation surrounding the fiber to accommodate for the lattice mismatch. Very limited or no tensile ductility was observed in them and therefore, these samples were deformed in compression to 0.5 % strain and 2 % strain. The low deformed 0.5 %-sample on TEM examination revealed dislocations only in the NiAl matrix. These dislocations were found to be <100> type which are believed to be emitted from the interface between matrix and fiber. At higher strains, dislocations in the Cr fiber were observed, although they did not observe any evidence of direct slip transfer across the interface.

Recently, Kwon et al. (Kwon et al., 2015) have discussed the dislocation evolution and deformation mechanisms in directionally solidified NiAl-Mo alloy. They have observed all three types of a<100> dislocations in the NiAl matrix in the undeformed specimens surrounding the Mo fiber, and the Mo fibers were predominantly dislocation free. This also corresponds to the observation by (Bei et al., 2007) where undeformed Mo fibers show very high strength values of ~9GPa, close to theoretical strength under compression, and were found to have very small number of dislocations (Phani et al., 2011). Based on the TEM observation of dislocation structures in undeformed, 4% strained and 16% strained samples, it is suggested that the NiAl matrix deforms first which leads to accumulation of <100> type dislocations on the interface. This activates a/2<111> slip in the Mo fibers with increasing strains and at very high strains, the a/2<111> slip in the Mo fibers then triggers the <111> type dislocations in the NiAl matrix.

2.3 Ductile phase toughening concept

Noebe and co-workers (Noebe, Misra, & Gibala, 1991) have explored the ductile phase toughening concept for NiAl based directionally solidified eutectics in detail in their work and proposed mechanisms involved both at macro- and at microstructural level, when a ductile phase is incorporated in the brittle matrix.

Macrostructural toughening

When a ductile phase is introduced in a brittle matrix, the ductile phase interacts with a crack growing through the brittle matrix. The ductile phase may be in the form of isolated particles, interpenetrating networks, or continuous lamellae or fibers. The amount of toughening observed is strongly related to the volume fraction and morphology of the second phase. Toughening can occur by several different mechanisms, such as crack bridging, crack blunting and crack deflection (Fig. 2.4).

Crack bridging (Fig. 2.4 a) helps in increasing the toughness of the brittle matrix by the ductile phase bridging the growing crack by requiring plastic deformation behind the crack front. Thus, further opening of the crack requires more energy. The volume fraction of the second phase and the size of the individual particles of the second phase have a great effect on the increase in toughness that can be achieved. Toughening by a continuous second phase is more powerful since the energy required for the plastic deformation of the bridging phase can become quite large. Limited debonding further enhances the toughening of the alloy, although extensive debonding is not favourable as this will lead to a reduction of the tearing modulus.

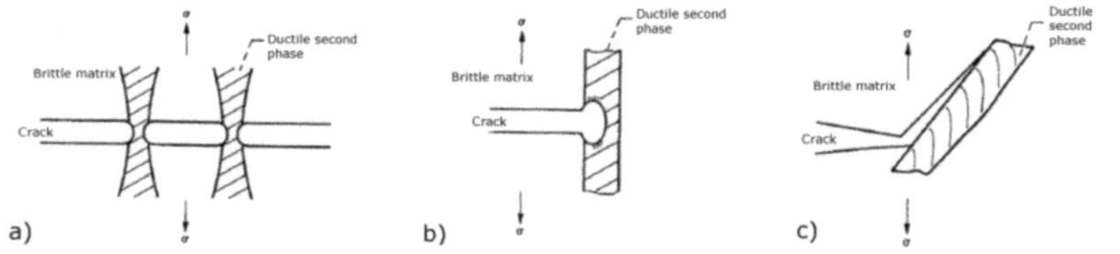

Figure 2.4: Macrostructural toughening mechanism in a brittle matrix containing a ductile phase; a) crack bridging, b) crack blunting and c) crack deflection (Noebe et al., 1991)

In crack blunting (Fig. 2.4 b), when the advancing crack tip meets the ductile phase, localized plastic deformation leads to the relaxation of the stresses at the crack tip, delaying the propagation of the crack. Low yield strength of the ductile phase will maximize this effect.

A second phase particle which has a weak bond with the matrix may lead to toughening by crack deflection mechanisms. As shown in Fig. 2.4 c, the crack redirects to a new path when it encounters a second phase which decreases the overall stress intensity at the interface. Crack deflection is the weakest of the three mechanisms. In general, more than one of these mechanisms can be involved in providing the toughening of the brittle matrix.

Microstructural toughening

At a microstructural level, toughening is achieved by successful transfer of plastic strain from the ductile phase to the brittle matrix. The process of strain transfer, in general, may take place either by direct transmission of dislocations from one phase to the other, or by nucleation of additional dislocations in the neighbouring phase.

Direct transmission of dislocations across the interface boundary requires a dislocation reaction, in which one part of the product dislocation is transferred to the second phase while the other part remains in the parent phase. While this has been observed in single phase materials across grain boundaries, it very difficult to achieve in two phase materials unless they have a coherent interface. Therefore, the strain transfer in the two-phase material is likely to take place by nucleating dislocations in the brittle phase.

Fig. 2.5 represents the schematic model of the strain transfer process in a two phase material which is based on the work by Lee and co-workers (Lee, Robertson, & Birnbaum, 1990), where they have examined the slip transfer process across grain boundaries in single phase materials. This understanding can be transferred to the slip transfer process in materials with a second phase as well. Three criteria were identified which determines the slip system which is activated in phase 2 (brittle) by stress concentration in phase 1 (ductile) due to the constrained deformation of phase 1 (ductile). When the orientation relationship between the two phases is such that the angle α is minimized and interface is strong, it leads to the rise in the stress at the interface due to the pile up of the dislocations in phase 1(ductile), which results in the dislocation nucleation at the interface. These dislocations which are generated at the interface, contribute to the plastic deformation of the phase 2 (brittle) matrix and thus providing additional ductility and toughness.

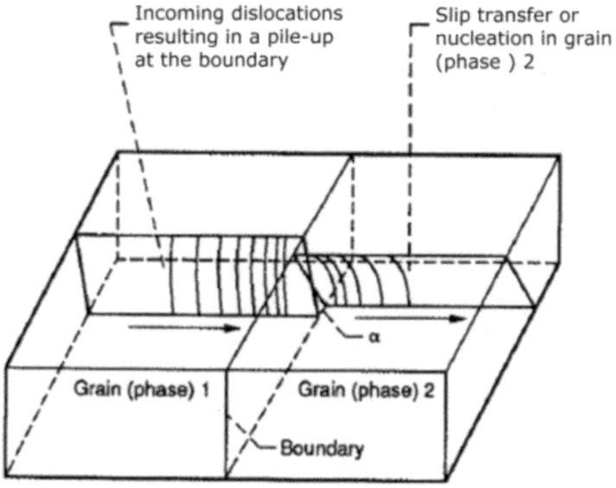

Figure 2.5: Schematic diagram of strain transfer process across the two phase interface ((Lee et al., 1990))

2.4 Small scale mechanical testing

As explained in the previous section, the mechanical properties of materials at the micro-/nano-meter regimes may be considerably different from the bulk properties which are usually determined by the conventional mechanical testing methods. Therefore, the need to determine the properties and response of materials at these small scales cannot be understated. Mechanical testing at small scales poses considerable challenges compared to conventional mechanical testing owing to the small size of the samples involved, where at least one of the dimensions is in the micro-/nano-meter regime, and lack of standardised testing. Application and measurement of very small quantities of load and displacements, proper imaging techniques with good resolution and ability to measure the properties free from the environmental 'noise' are very important to any small scale mechanical testing system in order to reveal and understand the deformation processes at this scale. In order to obtain clean and interpretable testing, several alternative testing methods have been developed in the recent past. Some of these developed testing methodologies, relevant to this study, are discussed in this section.

Nanoindentation

Nanoindentation is a versatile tool for studying deformation of material at small scale. Conventional indentation testing consists of contacting a material whose mechanical properties such as hardness and modulus are to be determined with a hard indenter tip, whose properties are known, by the application of a controlled load P. Nanoindentation is the extension of these methods on the micro and nano-scale regime where the depth of indentation is in nanometers rather than micrometers or millimeters. One of the important features of nanoindentation is the indirect measurement of the contact area. Due to very small size of the indents, optical methods to determine contact area for the hardness evaluation can lead to large errors in the measurements. The contact area in nanoindentation is determined by measuring the depth of indentation during loading and unloading and converting this into area using the known geometry of the indenter (Oliver & Pharr, 1992, 2004). This involves simultaneous measurement of the applied load P and the indentation depth h, down to nanometer regime. Therefore, these techniques are also sometimes referred to as depth sensing indentation technique or instrumented indentation technique (IIT) (Fischer-Cripps, 2004). A comprehensive review of nanoindentation method and its application can be found in (Palacio & Bhushan, 2013; Schuh, 2006). A typical modern nanoindentation set up consists of an indenter head (for the application of load by a coil/magnet assembly), an optical microscope for visualization of sample surface and indents, and an x–y–z motorized table for positioning and transporting the sample between the indenter assembly and optical microscope (Fig. 2.6).

One important feature of modern day nanoindentation testing is the use of the Continuous Stiffness Measurement (CSM) (Oliver & Pharr, 2004) which is used to measure the contact stiffness 'S', at each point during the loading. Traditionally S was determined by calculating the slope of the unloading cycle of a nanoindentation test. But this only enabled the users to calculate the stiffness at the maximum load and the amount of data used to determine the initial unloading stiffness was not standardized, which was a source of error in these measurements. CSM enables a continuous measurement of the stiffness by imposing a small dynamic oscillation on the force signal and measuring the amplitude and phase of the corresponding displacement signal by means of frequency specific amplifier.

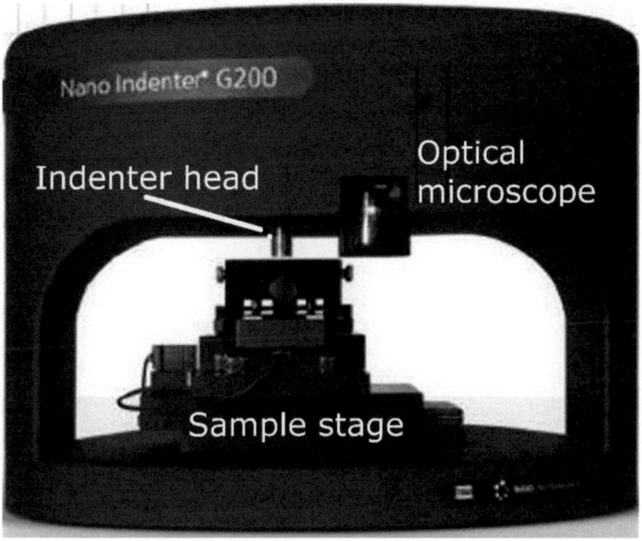

Figure 2.6: Nanoindenter G200 set up showing the indenter head, optical microscope and sample stage for transporting sample between indenter head and microscope

The distinct advantages of using CSM can be summarized as.

1. Provides continuous results as a function of depth.
2. Calibration time is heavily reduced as there is no need of multiple loading.
3. Effects of time dependent plasticity and thermal drift are reduced.
4. It also helps in identifying the point of initial contact as the stiffness changes suddenly at the point of initial contact.

Although, the most common use of nanoindentation is to determine the Hardness and Young's modulus, considerable advances have been made in the measurement of the other parameters such as hardening exponent, creep properties, residual stresses, nano tribological studies, fracture properties (Fischer-Cripps, 2004; Schuh, 2006).

Due to the very small size of the samples probed during a typical nanoindentation study, it has proved to be apt for detection and in-depth study of different discrete events occurring during the deformation of the material, for example detection and understanding plastic yield at nanoscale, development of dislocation network manifested in terms of pop-in events associated with movement, multiplication and evolution of complex defect structures, understanding mechanical instabilities occurring as burst of localized stains in amorphous materials and phase transformation under application of load (Fischer-Cripps, 2004).

In-situ tensile testing

Tensile testing has been the standard testing technique for bulk materials due to its ability to provide straightforward acquisition and interpretation of the important materials parameters such as yield strength and Young's modulus. Tensile testing is suitable for testing high aspect ratio quasi 1-D structures as it is not affected by elastic instabilities. In-situ testing of materials at small scale is an attractive technique as it allows for the observation of the underlying deformation mechanisms while monitoring the mechanical response, thus offering useful insights for establishing structure-property relationship.

The major challenges for any in-situ testing can be summarised as:

1. Fabrication of stress free and free-standing specimen
2. Proper gripping of the specimen
3. Accurate alignment of the specimen with the loading axis, and
4. Generation and detection of very small forces and displacements

Brenner in 1950s (Brenner, 1956, 1957) reported the pioneering work on tensile testing metallic filaments with diameters in the range of few micrometers using an instrumented tensile testing set-up. Qualitative stress-strain data was obtained using the optical microscope for imaging in the set-up. Since then, a variety of similar approaches has been used to characterize fibers, whiskers and filaments. Throughout the following decades several developments have been made to incorporate instrumented tensile testing into electron microscopes, which has allowed for testing of smaller fibers, in micro-/nano-meter regimes, which are too small for imaging by conventional light microscopy. Actuation and measurement of load and displacement are commonly achieved by piezoelectric, thermal, electrostatic and electromagnetic techniques. Development of Micro/ Nano electro mechanical systems (MEMS/ NEMS) have enabled the whole testing setups to be fitted inside SEM chambers and even into TEM sample holders (Kang, Rajagopalan, & Saif, 2010).

Yu et al. reported the use of Atomic Force Microscope (AFM) cantilevers and combination of commercially available stages for manipulation of the specimen while piezo actuator was used for displacement application. The AFM cantilevers act as a force sensor simultaneously. Zhu et al. (Y. Zhu & Espinosa, 2005) studied the size effect on mechanical behaviour for nanowires in TEM, using a MEMS based set-up. Kiener et al. studied FIB prepared miniature dog-bone shaped samples in-situ in TEM. A form locking grip was produced by FIB milling and attached to the head of a commercial nanoindenter for tensile testing. In-situ tensile testing using a combination of SEM and FIB (Rzepiejewska-Malyska et al., 2008) have sparked several studies in recent years. Gianola et al. (Gianola et al., 2011)reported the development of quantitative in-situ testing approach using a dual beam SEM/FIB system. Load application and displacement measurement was done by a nanomechanical transducer. Manipulation and alignment of the specimen was carried out by using custom build set-up capable of 3-axis positioning stage and nano-manipulator systems. SEM images captured at regular intervals were used for strain calculation by digital image correlation (DIC).

Micro pillar compression

Micro pillar compression testing was introduced by Uchic et al. in 2005 (Uchic & Dimiduk, 2005) for evaluating the flow properties of the materials using small scale specimens ranging from several micrometers to sub micro-meter in size. Micro pillars were prepared by Focused Ion Beam (FIB) milling into the bulk substrate. Thus, the substrate acts as the lower compression plate during the test, when the top surface of the pillar is compressed by a flat punch indenter tip in a nanoindenter system. A large amount of research has been dedicated since then following the technique to investigate a variety of materials at small scales. Several reviews of micro pillar compression testing, observed phenomenon and concepts explaining the observed phenomena has been published as well (Greer & De Hosson, 2011; Greer, Weinberger, & Cai, 2008; D. Kiener, Motz, & Dehm, 2009; Uchic, Shade, & Dimiduk, 2009a; T. T. Zhu, Bushby, & Dunstan, 2008).

The majority of micro pillars are prepared by FIB milling, which is useful in the careful selection and control of the location and the size of the micro pillar prepared. Fabrication of micro pillars by FIB milling can be classified into two main processes. Annular milling is the first process where the ion beam is oriented normal to the sample surface. The samples fabrication process is easy in this orientation and can be produced by stock milling patterns in a relatively short time. Single passes of low beam currents with a longer dwell time were used to remove the material from the rim of the pillar in steps till the desired diameter is achieved. One of the most important concern with the samples prepared by annular milling is that there is always a certain degree of taper associated with the process such that the diameter at the top is always smaller than the base diameter. This, in general, leads to the concentration of plastic deformation on the top of the pillars and induced a strain gradient along the pillar. Another concern in the annular milling is the difficulty to control the aspect ratio (length to diameter ratio) of the pillars. In general, the preferred taper in the micro pillars prepared by annular milling is within 2 to 5 degrees and the preferred aspect ratio is in the range of 2:1 to 3:1.

In lathe milling process, the sample is oriented in such a way that the angle between the sample surface and FIB column is minimum. The milling is done on the sides of the pillars along the length. After each milling step, the pillar is rotated few degrees and then the same process is repeated. The whole process is usually automated, a special fiducial mark is milled on the top of the pillar so that the software can take images after each milling step and realign the pillar automatically. The advantages of using this process is that it produces very smooth and taper free pillars. On the other hand, due to the direct exposure of sample under the ion beam and the requirement to take several FIB images and fiducial marks, the chances of Ga ion implantation in the sample are very high which can alter the properties measured. Also, the whole process is fairly cumbersome and difficult to control.

Irrespective to the two process used, a common concern over samples prepared by the FIB milling is the effect of irradiation damage due to high accelerated Ga^+ ions on the mechanical properties of the materials evaluated by micro pillar compression tests. The work by Greer and Nix (Greer & Nix, 2006; Greer, Oliver, & Nix, 2005) evaluated the effect of FIB irradiation damage by comparing samples with varying degree of irradiation damage. Their experiments reveal that the all the samples, prepared by FIB milling having irradiation damage, showed similar properties, while electrodeposited micro samples had higher strengths than those with FIB damage, although both type of processes showed similar size dependent strengthening trends. Kiener et al. (Daniel Kiener, Motz, Schobert, Jenko, & Dehm, 2006) argued that the Ga^+ ion contamination in the samples prepared by the FIB milling can act as a barrier for dislocation motion which can influence the mechanical response of the material. Hütsch and Lilleodden (Hütsch & Lilleodden, 2014) have compared the effects of ion damage during the FIB milling of pillars in lathe and annular milling processes and found that although the lathe milling process produces better samples geometries, the amount of ion damage is much larger compared to the Annular milling and may affect the properties of the material under investigation significantly.

Other methods of preparing micro pillars include microelectronic-based processes to create array of metal pillars (Greer & Nix, 2006) or semiconductor pillars (Moser, Wasmer, Barbieri, & Michler, 2007) and selective etching of directionally solidified alloys (Bei et al., 2007). These studies on the micro pillars prepared by alternative methods not only highlight the effects of FIB damage on the samples, but also provide useful insight into the size dependence of strength observed in these tests.

2.5 Mechanical size effect

Mechanical size effect can be described as the change in response of a material due to the reduction in the length scale of either the internal structure of the material or the reduction in the overall dimension of the sample itself. Classical continuum models typically used to describe the material behavior at bulk scale are size independent. However, it has been shown repeatedly over the last decades, that when the materials dimensions are in the range of micro/ nano meter regime, the strength increases rapidly and conventional models are not suitable to describe the material behavior at these scales. The size effect in material can be classified into two broad categories, 'intrinsic' and 'extrinsic' size effects. Intrinsic size effects result due to the changes at microstructural level, while the extrinsic size effects arise from dimensional constraint due to small sample size or even small test volumes under strain. There are several comprehensive reviews available in the literature focusing on understanding mechanical size effects in metallic materials (Greer & De Hosson, 2011; T. T. Zhu et al., 2008). The mechanisms and results discussed in the present work are focused on the confined deformation of single crystals and therefore, extrinsic constraints. The extrinsic size effect is discussed in detail in this section, although a brief overview of the intrinsic size effect is presented as well.

Intrinsic size effect

Intrinsic size effects are primarily controlled by materials processing history. Grain size, particles, solute atoms, precipitates and dislocations act as a microstructural constraint in the classical metallurgical concepts. Interaction of dislocations with these barriers lead to the length-scale dependent behaviour of the materials. Particles, solute atom and dislocation spacing and grain size with respect to the dislocation loop diameter are the relevant parameters for the occurrence of the size effect in mechanical behaviour (Arzt, 1998). For the latter, the dependence is given in general by a power-law relationship:

$$\sigma_y = \sigma_0 + k\, d^n \tag{2.1}$$

where, σ_y is the yield strength and σ_0 is the bulk strength, k is a constant, d is the grain size of the investigated material and n is the power-law coefficient.

In a polycrystalline material, the interaction between the grain boundaries and the defects is the main cause of the observed size effect as first reported by Hall (Hall, 1951) and Petch (Petch, 1953). The above relationship with power-law coefficient of -0.5 is well known as 'Hall-Petch relationship' which describes the increase in the yield strength of a polycrystalline material with the decrease in grain size, as the grain boundaries in these polycrystalline materials act as obstacles to the dislocation motion. Although traditionally the increase in strength was attributed to dislocation pile-up on the grain boundaries, other models namely, grain boundary source model (Li, 1963), slip distance model (Chia, Jung, & Conrad, 2005) and geometrically necessary dislocation (GND) model (Ashby, 1970) have also been proposed to describe the size effect on strength (T. T. Zhu et al., 2008).

When the grain size of the nanostructured materials are reduced in the range of ~40 nm, alternate plastic deformation mechanisms take over the traditional mechanism such as grain boundary (GB) sliding, partial dislocation emission and adsorption at grain boundaries, since the nano-sized grains cannot accommodate multiple lattice dislocations (Greer & De Hosson, 2011). Upon further reduction of grain size to ~20 nm, a saturation in strength or even an 'inverse Hall-Petch' relation has been observed due to the activation of the grain boundary assisted deformation. The dependence of strength of the materials with respect to grain size is shown in the figure 2.7.

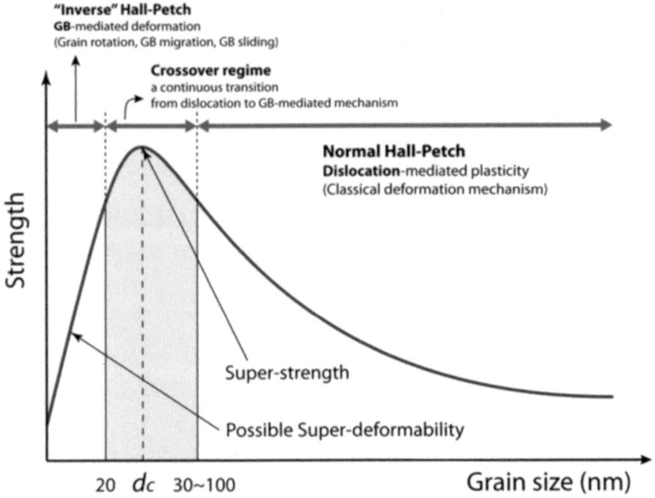

Figure 2.7: Strength of polycrystalline materials with respect to the grain size: Hall-Petch relationship and transition to inverse Hall-Petch (Greer & De Hosson, 2011)

Extrinsic size effect

Extrinsic size effects arise due to dimensional constraint in the sample structures where one dimensions is very small compared to the other two, such as thin films and nano fibers. But, examination of pure dimensional constraint effects may be affected if the microstructural length scale is in the range or below the thickness of the structure. Therefore, to study the effect of dimensional constraint only, the preferred test geometries are single crystalline pillars and micro/nano fibers and ribbons.

The mechanical properties of thin films have attracted numerous studies and a variety of test methods have been developed in the past decades (Hommel, Kraft, & Arzt, 2017; O. Kraft, Hommel, & Arzt, 2000; O. Kraft & Volkert, 2001; Shen et al., 1998). Experiments carried out on the thin films on bulk substrate under mechanical and thermal loading have all revealed that the plastic response depends strongly on the thickness of the thin films with yield strength has been found to be inversely proportional to the film thickness. Misfit dislocation channelling theory, proposed by Frank and van der Merwe (Frank & van der Merwe, 1949), has been used to explain the increase in strength with decreasing film thickness, assuming that the misfit dislocations are generated from the substrate and passed along the thin film channel. Plastic deformation occurs when the stresses are high enough to bow the dislocation loop such that it fits in the thin film channel. For thin films with thickness in the micro-meter regime, the strain gradient plasticity theory (Hutchinson, 2000) is also used to explain the strengthening mechanism. The yield strength increase is believed to be caused by the presence of the GNDs (Ashby, 1970) which appear to accommodate the strain gradient in the thin films. Gruber et al. (Gruber, Solenthaler, Arzt, & Spolenak, 2008) have explored the deformation of thin film under one dimensional constraint using Au thin films on a compliant substrate by a synchrotron-based tensile testing technique. They found that the flow stress of the films increases with the decreasing thickness but for films thickness below 80 nm, constant and even decreasing flow stresses were observed with decreasing thicknesses. The change in behaviour is attributed to the change in the deformation mechanism from perfect dislocations to partial dislocations as carrier for plastic strain.

In the past decade, there have been numerous studies, demonstrating the significant increase in strength associated with the reduction in the sample size for single crystals at micron- and sub-micron length scales. The development of instrumented test equipment with highly sensitive force and displacement measurement and advanced imaging techniques had led to the wide variety of testing techniques that have been developed to explore the extrinsic size effect on the properties of the materials. Several reviews (Greer & De Hosson, 2011; Oliver Kraft, Gruber, Mönig, & Weygand, 2010; Uchic, Shade, & Dimiduk, 2009b; T. T. Zhu et al., 2008) provide a comprehensive review of the dimensional constraints in materials and their influence on the properties of the materials.

One of the earliest research of the properties of small scale single crystal was presented by Brenner (Brenner, 1956, 1957), where he showed that the single crystalline filaments, termed as whiskers, were able to withstand unprecedented levels of stresses and strains with a lot of them approaching the theoretical strengths of the material. These results firmly established the importance of specimen size for evaluating the strength of the components. These whiskers, although showing considerable scatter, showed sharp yield points followed by extensive glide at stresses lower than the yield stress. When the fractured whiskers were re-stressed, they showed even higher stresses, suggesting that deformation in these whiskers is dominated by dislocation nucleation and followed the weakest link approach. Furthermore, the strength of the whiskers were also found to be inversely proportional to the length of the whiskers tested, which suggested that the longer the specimen, the more is the probability to find a statistically stored dislocation which also explained the scatter in the data.

More recently, tensile testing small whiskers was extended to nano meter regime when tensile tests were performed on Cu nanowires (Richter et al., 2009), which reached even higher stresses than reported by Brenner. These nanowires were grown by novel synthesis techniques resulting in single crystalline, high aspect ratio and defect-free nanowires, confirmed by combination of different electron microscopy techniques. In-situ tensile tests were performed on these Cu nanowires between sizes of 75 nm and 300 nm inside a dual beam SEM/FIB system. The thicker wires (>200nm) showed fracture occurred on slip planes having highest resolved shear stresses suggesting highly localized dislocation activity, while the thinner nanowires showed brittle fracture in absence of dislocations. No plastic necking was observed in either case. The work by

Johanns et al. (Johanns et al., 2012) tested high aspect ratio Mo rich single crystal fibers in tension and compared the mechanical behaviour with low aspect ratio micro pillar in compression (Bei et al., 2007). They show that although the smallest micro pillar in compression reach theoretical strength values, the fibers with similar diameters show a large scatter in the strength values between 1 GPa and 10 GPa. This behaviour can be explained by a weakest-link concept, which suggests that a large volume of the probed material will have a high probability of containing initial defects, which would lead to localization of plasticity, thereby reducing the strength. Theoretical (Senger, Weygand, Motz, Gumbsch, & Kraft, 2011) and experimental (Mompiou et al., 2012) studies have been performed to confirm the weakest link approach. Senger et al. have shown by 3D Discrete Dislocation Dynamic study that flow stresses in small samples follow weakest link statistics and results of simulations matches the experimental observation well. Mompiou et al. have used in-situ SEM and TEM experiments to directly link the size effect mechanical behaviour of sub-micron Al fibers with the initial dislocation density and distance between the source and the surface.

Micro scale tensile tests on the samples produced by focussed ion beam (FIB) milling were performed by Kiener et al. (D. Kiener, Grosinger, Dehm, & Pippan, 2008) on single crystalline Cu samples of diameters varying from 0.5 μm to 8 μm and also with varying aspect ratios between 1:1 to 13.5:1. Sample with high aspect ratios (>5:1) were reported to deform by slip in only one slip system. The presence of single slip indicated that the plastic deformation in these samples with higher aspect ratios is controlled by a number of single dislocation sources. Each source is responsible for certain plastic deformation followed by the activation of the next source. Although the strengths observed were much lower than the theoretical strengths, a weak size effect was observed which was attributed to the size of the dislocation sources governing plastic deformation. In case of low aspect ratio samples, a strong size effect was observed with respect to the samples size. In addition, significant hardening was observed in these samples when compared to the high aspect ratio samples. This is explained using the dislocation pile-up theory.

The study by Uchic et al. (Uchic, Dimiduk, Florando, & Nix, 2004) which showed the influence of diameter on the mechanical response of the micro pillars in different materials, was the first of many micro pillar studies carried out over the past decade. It has been observed that in general the strength of the micro pillars increases with decrease in the pillar diameter, showing stochastic behaviour. This is mainly due to the individual slip events during deformation and is strongly dependent on the initial defect structure. Since then, pillar compression has been one of the most popular test methods to evaluate strengths for a variety of materials and a wide range of diameters, ranging from tens of micrometers down to few hundred nanometers. In general, micro pillars show a strong size effect with respect to the pillar diameter with yield strengths found to scale according to eq. (1) with exponents between -0.5 and -1 (Reference).

Efforts to explain the strong size effect observed in pillar compression tests have resulted in various mechanisms being proposed over the years. Generally, *dislocation multiplication* is the main mechanism for strain hardening in metals as the increase in dislocation density leads to a hindrance of the movement of mobile dislocations through the dislocation forest (Taylor hardening (G. I. Taylor, 1934). For the small pillars, however, it has been suggested (Greer & Nix, 2006) that the dislocations may run out through the surface and thus the increase in strength is related to *dislocation starvation* in the micro pillars(Greer et al., 2005). They have argued that for small enough sample dimensions, the mobile dislocations can only travel small distances and will leave the sample at the surface of the sample instead of interacting and multiplying. Thus, to continue the plastic deformation, higher stresses are required to activate less favourable dislocation sources or to nucleate new dislocations in the sample, similar to the whiskers and nanowires. In-situ TEM study by Shan et al. (Shan, Mishra, Syed Asif, Warren, & Minor, 2008) provided evidence for dislocation starvation mechanism, when they observed that the high dislocation density is reduced by application of mechanical stresses. However, Kiener and co-workers (D. Kiener, Motz, Schöberl, Jenko, & Dehm, 2006) have argued that the size effect in micropillars can be attributed to some extent to FIB induced damage layers, which may cause *dislocation pile-ups* at boundary layers which hinders the dislocation movement and thereby causing strengthening. The important role of the initial dislocation structure for the strength of micropillars was highlighted by Bei and co-workers (Bei, Shim, Pharr, & George, 2008). They showed that similar to the study by (Mompiou et al., 2012) the amount of initial dislocation density greatly affects the mechanical response of the material. Investigation of pre-deformed pillars etched from directionally solidified NiAl-Mo eutectic alloy shows a transition from the bulk like behaviour at high pre-strains (11%) to intermittent stochastic behaviour at low pre-strains (4%) to high strength values approaching the theoretical strengths for undeformed (0%) samples. Furthermore, these results are in contrast to other pillar experiments as the samples sizes affects the pillar strengths much less than the initial dislocation density (Figure 2.8). Since the pillars were produced from an alternate method, they were free from Ga^{+} implantation from FIB and therefore, also highlights the importance of FIB beam damage on the mechanical response of the materials.

None of the above proposed models are capable of explaining the size effect due to the dimensional constraint solely. Different boundary conditions dominate which mechanism dominates the deformation process and very often, various mechanisms may co-exist and interact with each other.

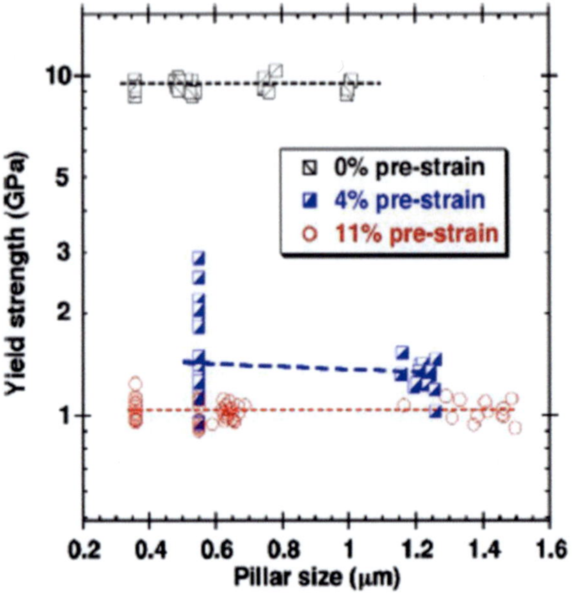

Figure 2.8: Yield strength of pillars with different prestrain levels with respect to the diameter of the pillars *(Bei et al., 2008)*

Size effect in BCC metals

Plasticity in bcc metals, in general, is more complex as compared to fcc metals and deformation is influenced by temperature, strain rate and orientation. Several studies have focused on understanding the small scale mechanical response of single crystalline bcc metals, many of them on Mo. Brinckmann et al (Brinckmann, Kim, & Greer, 2008) have performed compression experiments on Mo and Au nano pillars and found that both metals exhibit size effects along with stochastic discrete bursts in the stress-strain curves, although the strengthening behaviour and the fraction of theoretical strength reached was found to be drastically different where 40 % of theoretical strength was achieved for Au nano pillars as compared to only 7% for Mo. This difference was attributed to fundamental difference in plasticity mechanisms in bcc and fcc materials, where for fcc, the mobile dislocations escape the crystal at the free surface, while for bcc, a dislocation multiplication mechanism is discussed where a dislocation loop forms kinks which generate other dislocation segments (Greer et al., 2008; Weinberger & Cai, 2008).

Kim et al. (Kim & Greer, 2009; Kim, Jang, & Greer, 2012) have also reported size dependent strength of (0 0 1) oriented Mo nano pillars in tension and compression and showed that higher flow stresses in smaller samples is due to an increase in yield strength rather than the strain hardening or other mechanisms. They also presented TEM micrographs of nano pillars before and after the test revealing the formation of entangled dislocation sub-structure, which supports the view of different deformation mechanism in bcc as compared to fcc. Micro-compression experiments on different oriented Mo pillars by Schneider et al. (Schneider, Kaufmann, et al., 2009; Schneider, Clark, Frick, Gruber, & Arzt, 2009) revealed that the power law exponent (Eq. (1)) is dependent on the orientation, -0.22 for [001] and -0.34 for [220]. The pillars were also loaded with varying strain rate and the authors report a strong strain rate dependence for pillars in both orientations with activation volumes only marginally smaller than in the bulk.

Besides Mo, Schneider et al. (Schneider, Kaufmann, et al., 2009) studied the compression of Mo, Ta, W and Nb where they observed that the power law exponent correlates with the inverse critical temperature T_c of each material. This may be attributed to the low mobility of screw dislocations below T_c. The authors also report that the strength difference among these materials is lowest for the minimum diameter of 200nm, implying that as the pillar size decreases the Peierls mechanism becomes less dominant. Kim et al. (Kim & Greer, 2009) also studies the same four metals under tension and compression. They report a strong size effect, with power law exponent for size dependent strength for Nb being much higher than the other three. They also observed significant tension-compression asymmetry which is itself size dependent. Han et al. (Han, Bozorg-Grayeli, Groves, & Nix, 2010) investigated compressive properties of V nano pillars and reported the power law strengthening exponent of -0.79, which is consistent with other small scale studies on bcc metals.

2.6 Helmholtz research school 'IMD'

The development of novel alloys requires a lot of effort from different fields of material science to build the required knowledge base. The Helmholtz Research School on *"Integrated materials development for novel high temperature alloy"* or "IMD", funded by the Helmholtz Association Germany, provides such as platform for a knowledge-based and research-oriented materials development.

Directionally solidified eutectics provide a promising approach for development of novel alloys for high temperature structural applications particularity for improving creep resistance and ductility of NiAl based alloys. As a starting step, the NiAl-Cr directionally solidified eutectic alloy has been selected since Cr has lowest density among all the refractory metals available for alloying with NiAl to produce directionally solidified eutectics. The graduate school is structured to have eight research areas comprising of processing, microstructure, properties and modelling/simulations which work in close collaboration with each other in order to strengthen the scientific basis in materials research for energy applications and related technologies, leading to an integrated materials development effort (Fig. 2.9). The different institutes and the respective research areas participating in the IMD are listed in table 2.2.

This project belongs to research area four, RA IV, in the framework of IMD with the focus on multiscale mechanical characterization of directionally solidified eutectics in order to develop a mechanistic understanding of the deformation processes and toughness at room temperature.

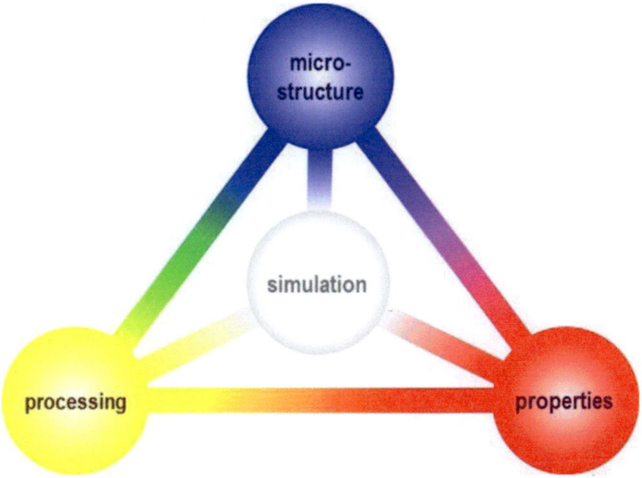

Figure 2.9: Schematic diagram of the cross linking of different research areas within the framework of IMD

Table 2.2: Research areas and participating institutes of IMD

	Research Area (RA)	Participating institute
I	Thermodynamics	IAM-AWP, KIT
II	Directional Solidification	IAM-WK, KIT
III	Phase field simulation	IAM-ZBS, KIT
IV	Multiscale mechanical testing	IAM-WBM, KIT
V	Microstructure and Creep deformation mechanism	IAM-AWP, KIT
VI	Load partitioning and internal damage	IAM-WK, KIT
VII	Discrete dislocation dynamics	IAM-ZBS, KIT
VIII	FEM simulation of thermomechanical behaviour	ITM, KIT

2.7 Open questions and aims of the study

Despite NiAl having very attractive set of properties for high temperature structural application, its use in these application is inhibited by:

i) Low ductility and fracture toughness at room temperature
ii) Poor creep properties at elevated temperatures.

A lot of research has been dedicated towards improvement of NiAl with respect to drawbacks mentioned above and has been reviewed. Directional solidification of eutectics of NiAl, alloyed with refractory metals to produce an in-situ composite appears to be the most promising route and several studies indicate significant improvements in ductility and fracture toughness. The most important parameters with respect to the directional solidification of these eutectics are the species and amount of alloying additions to produce the eutectic and the solidification speed for the directional solidification. Both factors, have a huge effect on the final microstructure of the alloy, as has been discussed previously. With carefully chosen alloying addition and solidification parameters, a variety of microstructures can be obtained such as fibers, plates or lamellae. The final microstructure then in turn affects the properties of the alloys prepared.

The research in the past, although having reported significant improvements over binary NiAl, has not obtained a consensus on the optimum alloying additions and processing conditions to be used for the best properties. Probably, this is due to the fact that in all the research, the alloying additions and processing conditions seems to be chosen randomly and a systematic research on these alloys has not been carried out. Furthermore, till date, though some studies have been performed exploring the mechanisms behind the improvement in properties such as crack bridging, crack blunting and crack deflection, a clear understanding of the deformation mechanisms involved in the improvement in the ductility and fracture toughness has not been developed.

Another important factor in the development of understanding of the deformation mechanisms is that in the directionally solidified eutectics, the characteristic length scales which are the fiber diameter/lamellae thickness, fiber/lamella spacing, are in general in the micro/nano meter regime. Considering the strong mechanical size effects observed in the variety of materials when the size of the components falls in the micro/nano regime, it is very important to examine these materials on the relevant length scales to fully understand the deformation mechanisms and the role of individual phases and interfaces between the two phases on the overall mechanical properties of these eutectic alloys. Based on these insights, an effort can be made to establish the optimum alloying additions and processing conditions in order to achieve the best possible properties.

Therefore, the main aims of this thesis were identified as:

- To characterize the mechanical behaviour of directionally solidified NiAl-Cr eutectics, at varying length scales, addressing fiber, matrix and interfaces, in order to fully understand the role of these individual phases and interfaces on the overall mechanical properties.
- To develop a mechanistic understanding of the role of processing conditions on the mechanical properties of the directionally solidified eutectics by examining the alloys processed at different solidification speeds.

3 Experimental details

3.1 Material: NiAl-Cr directionally solidified eutectic

The material for this project was the directionally solidified eutectic alloy in the quasi-binary system NiAl-Cr. Chromium was selected as the alloying element due to its high oxidation resistance and lowest density among all the refractory metals identified to be alloyed with NiAl to improve its properties. The composition of the eutectic is well defined in the literature (Walter & Cline, 1970, 1973) as Ni- 33%Al- 34% Cr (at. %). The material with this composition is cast by arc melting into rods. These cast rods are then processed by directional solidification in a zone melting device to produce the directionally solidified alloys with a highly aligned microstructure (Walter & Cline, 1970)(Haenschke et al., 2010; Rablbauer et al., 2004; Walter & Cline, 1970, 1973; Yang, 1997b). The samples were produced and directionally solidified with different solidification speeds (i.e. 20, 50 and 80 mm/h) at Oak Ridge National Laboratory, U.S.A. The samples used in this study are termed as ORNL_xx, with xx being the solidification speed of the samples.

3.2 Microstructure analysis:

3.2.1 Sample preparation

The as received material was in the shape of rods with about 10 mm in diameter and several centimeters in length. Out of these rods, button shaped samples of about 4-5 mm were cut out by machining. In addition,

3-4 mm slices from the longitudinal direction of the samples were also cut out in order to evaluate the longitudinal section as well. The samples were then cold mounted using the Demotec15® powder and binder. The cold mounted samples were ground and polished using standard techniques up to 0.25 µm diamond paste to achieve a mirror like surface. After grinding and polishing, the samples were then broken out of the mold and samples and transferred on the Scanning electron microscope (SEM) sample holders for microstructure examination using a dual beam SEM/FIB system (Nova Nanolab 200, FEI, Hillsboro, Oregon 97124, USA). These samples on the SEM sample holders were also used for the sample fabrication using Focused Ion Beam (FIB) explained later in this chapter.

3.2.2 Dual beam SEM/FIB

Scanning electron microscopy combined with focused ion beam milling is one of the most important techniques used in recent years as a characterization and sample fabrication tool at micro and nano scale. In this work as well, owing to the sub-micron features, microstructure examination of the alloy is done by a dual beam SEM/FIB system, (Nova Nanolab 200, FEI, Hillsboro, Oregon 97124, USA). It consists of a SEM column along with a Ga^+ ion column which is mounted at an angle of 52° with respect to electron column. The electron column is equipped with a thermal field emission gun, which has several acceleration voltages, ranging from 1 KV to 20 KV. Usually an accelerating voltage of 5 KV and a beam current of 1.6 nA was used for microstructural examination in this study. The electron beam has a maximum resolution of ~1 nm.

In FIB, Ga^+ ions are emitted from the Ga^+ ion source and are accelerated towards the sample by applying a voltage 30 KV. The Ga^+ ions penetrate the sample and the electrons generated can then be used for imaging. The imaging resolution is about 7 nm but is reduced sharply when the beam current is increased. In addition, the focused ion beam, due to the heavy Ga^+ ions, sputters atoms from the surface away. The rate of sputtering depends on the sample material, accelerating voltage and beam current. Depending on the application, the ion beam current can be chosen to be between 10 pA, for imaging and 20 nA for large milling rates. In this study, FIB milling has been used extensively for microstructure examination, and preparation of micro-pillars and micro scale double cantilever beam samples. A schematic diagram of a dual beam SEM/FIB is depicted in Fig. 3.1 along with the photograph of the SEM chamber.

3.3 Nanoindentation

Nanoindentation was used in this study to evaluate and compare the hardness and Young's modulus of directionally solidified eutectics prepared at different solidification speeds. The nanoindenter used in this study is a Nanoindenter G200 from Keysight Technologies GmbH, Böblingen, Germany. A review of the basics of nanoindentation and its applications has already been presented in section 2.5. The testing and analysis method used in this study is explained in detail in this section.

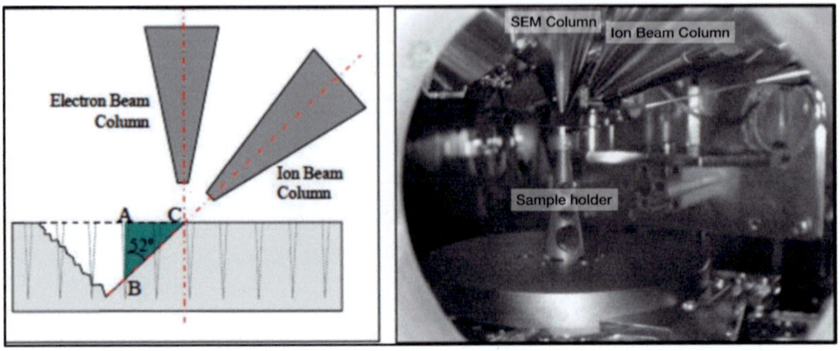

Figure 3.1: (a) Schematic diagram of a dual beam SEM/FIB and (b) photograph of the SEM chamber showing the arrangement of the columns. The angle between the SEM column and FIB column is 52°

A three sided pyramid Berkovich tip was used. Due to relatively small size of the indents, which were about 5 um wide, it was possible to use the same sample prepared for the microstructural examination for the nanoindentation study as well as the pillar fabrication process. For the nanoindentation study, special mounting blocks were prepared so that the sample on the SEM holders could be used on the nanoindenter sample stage without the need to transfer the sample between SEM sample holder and nanoindenter sample holder every time. At least 20 indents on each sample to get the average value of Hardness H, and Young's modulus E. All the tests were carried out using the continuous stiffness measurements (CSM) option which means the contact stiffness of the tip with the sample is determined continuously throughout the measurement.

3.3.1 Test procedure

The test begins with the loading section where the tip is brought in the contact with the sample surface. The CSM option is helpful here in identifying the exact point of initial contact. This is crucial for identifying the exact point of origin of the load-displacement curve, for which the data can be corrected even after the experiment. After the initial contact, the tip is then loaded till the maximum depth is reached. The strain rate for the loading for all the samples was 0.05/s. The maximum depth of indentation was chosen to be 2000 nm, such that the indentation area was large enough that multiple fibers were deformed under the contact area, in order to measure the average values for H and E, possibly close to the bulk properties. After the maximum depth was reached, the load at this depth was held constant for 10 s after which sample was unloaded to 10% of the maximum load, where it was again held constant for 300 s to measure the thermal drift in the system. Finally, the sample was completely unloaded. A typical load vs displacement curve obtained in such test is shown in Fig. 3.2.

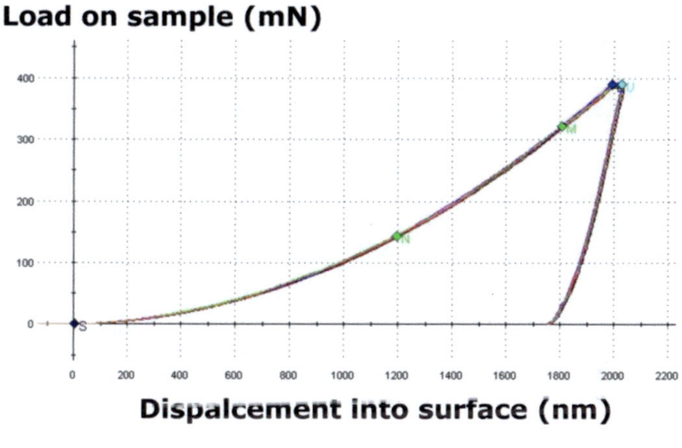

Figure 3.2: Typical load vs displacement curve obtained in a typical nanoindentation test to a maximum displacement of 2000 nm.

3.3.2 Data analysis

Using the Nanosuit® software (Keysight Technologies GmbH, Germany), the load vs displacement curve is obtained from the raw test data. These data were then analysed using the Oliver and Pharr method (Oliver & Pharr, 1992, 2004). The three basic quantities required for determination of hardness and modulus are load, P, displacement, *h*, and contact stiffness, S. The ratio of load over stiffness squared, P/S^2, is a direct measurable parameter which is independent of the contact area assuming the hardness and modulus do not vary with depth.

$$\frac{P}{S^2} = \frac{\pi}{(2\beta)^2} * \frac{H}{E^2} \tag{3.1}$$

The area independence of P/S^2 allows the determination of the load frame compliance, C_f. The total measured compliance C is composed of the load frame compliance, C_f, and the specimen compliance, C_s, acting as two springs in series.

$$C = C_f + C_s \tag{3.2}$$

The basic principle for load frame compliance measurement is that at depths greater than few hundred nanometers, where the hardness and modulus should independent of the depth, P/S^2 should be constant. Thus, the proper value of load frame compliance is obtained by changing the value of C_f for which a flat P/S^2 versus *h* curve is obtained for larger depths.

With load frame compliance and indenter area function known, hardness and modulus can be evaluated as:

$$H = \frac{P}{A} \tag{3.3}$$

$$S = \beta * \frac{2}{\sqrt{\pi}} E_{eff} \sqrt{A} \tag{3.4}$$

where E_{eff} is the effective modulus defined by

$$\frac{1}{E_{eff}} = \frac{1-\vartheta^2}{E} + \frac{1-\vartheta_i^2}{E_i} \tag{3.5}$$

E_i and E are the Young's modulus of the indenter and specimen respectively and ϑ_i and ϑ are the Poisson's ratio of the indenter and the specimen respectively. When the Poisson's ratio of the material evaluated is not known, the modulus is obtained by assuming the Poisson's ratio as 0, and it is reported as "indentation modulus".

The first step is to identify the correct point of initial contact for each test. This is done by extracting the phase angle vs displacement curve as shown in Fig. 3.3, where the point of contact is identified as sharp change in the phase angle with respect to the displacement. In the figure, the sharp change in the phase angle with respect to displacement in surface is marked by symbol S marking the detection of surface by indenter tip.

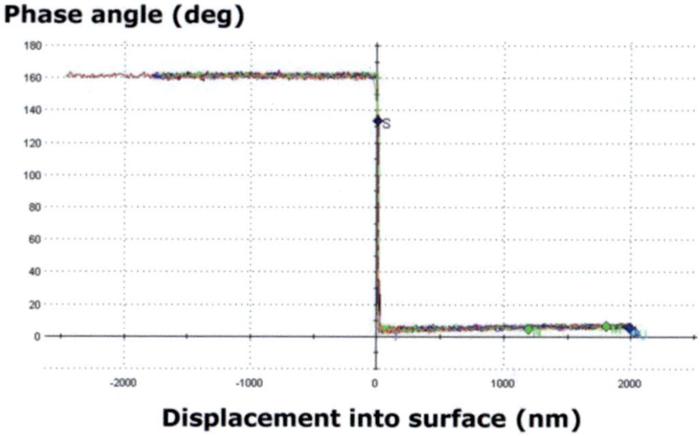

Figure 3.3: Phase angle vs displacement plot of a nanoindentation test, used to identify the point of initial contact of the tip with the surface S

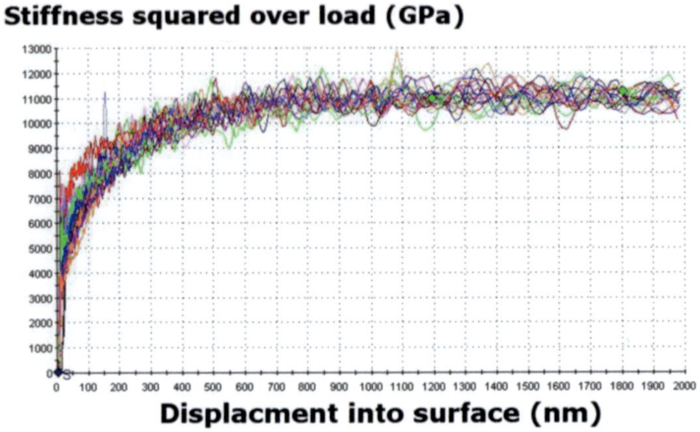

Figure 3.4: Stiffness squared over load vs displacement plot obtained for nanoindentation tests. The frame compliance is adjusted to obtain a flat trend at large depth

The next step is to obtain the stiffness squared over load (S^2/P) vs displacement curve and adjust the compliance value such that S^2/P does not change with the depth in the later part of the test (Fig. 3.4). This approach is based on the assumption that at depths larger than few hundred nanometers, the hardness and Young's modulus values do not change with depth and are free from any contribution from the machine and tip compliance, as explained in section 2.5.1. Once a flat S^2/P curve is obtained by adjusting the frame compliance C_f, H and E vs depth curves can be obtained. Since, the CSM option calculates the stiffness values at every depth, a continuous hardness and modulus curve is obtained and although H and E do not change with the depth, there are always some inherent fluctuations in the curve. Therefore, for all the tests, the values of H and E from a depth of 1200 nm to 1800 nm were averaged to get the H and E for the particular tests (Fig. 3.5). The values from all the tests from the same sample were then averaged to get the average value of hardness and modulus for that particular sample. The data for poisson's ratio for NiAl-Cr was not available in the literature. Therefore, a poisson's ratio of 0.3, as for most metals, was assumed to calculate Young's modulus. The input parameters used for the testing is tabulated in table 3.1.

Table 3.1: Important input parameters used for nanoindentation testing in this study

Input parameter	Value	Unit
Strain rate	0.05	1/s
Surface approach velocity	10	nm/s
Frequency	45	Hz
Surface approach distacnce	1000	nm
Depth limit	2000	nm

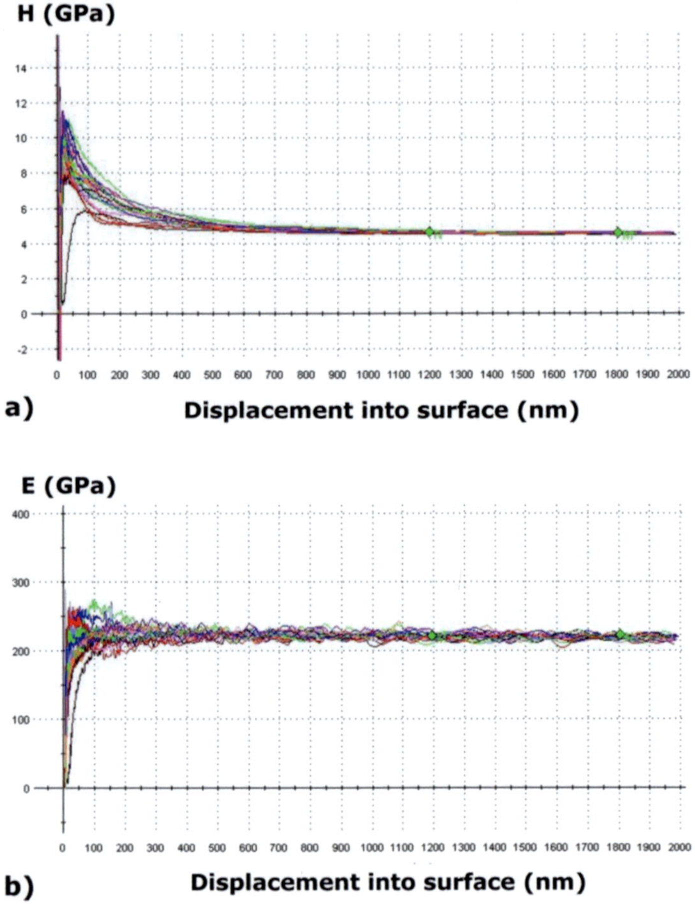

Figure 3.5: (a) H and (b) E vs the displacement into the surface plot obtained from nanoindentation tests, the data from 1200 to 1800 nm was used to get the average values for H and E for each test.

3.4 SEM in-situ tensile test

An important aspect of the mechanical characterization of the directionally solidified eutectics is to characterize the individual components. To characterize the individual Cr fibers, the technique for testing small test structures developed by D. Gianola and co-workers (Gianola et al., 2011) was employed. The tests were carried out by Charlotte Ensslen (IAM, KIT, Germany). This technique in which individual fibers can be tested in-situ inside an SEM is very helpful not only to test sub-micron structures but it also allows for direct observation and recording of the deformation process, which helps in developing our understanding of the mechanisms involved in the deformation (Gianola & Eberl, 2009; Johanns et al., 2012; Legros, Gianola, & Motz, 2010; Sedlmayr et al., 2012). The procedure for extracting individual Cr fibers from the test sample, a brief description of the test set-up and the test procedure is described in this section.

3.4.1 Extraction of Cr fibers

The first step of characterizing individual Cr fibers is to extract them from the as directionally solidified eutectic composites. For this, samples in the longitudinal sections were prepared from the samples produced at different solidification speeds. These samples were then cold mounted in epoxy resin and ground and polished in a similar manner as described in the section 3.2.1 to prepare a smooth surface free from defects and irregularities. After the surface preparation, the samples were subjected to chemical etching to remove the NiAl matrix. For this, a freshly mixed solution of 100 ml H_2O, 40 ml HCl (40% conc.) and 5gm CrO_3 was prepared. The optimum time of etching in order to remove enough NiAl was identified by trial and error to be of 30 s, resulting in the removal of about 1-2 mm of NiAl into the surface, which was found to be enough for the extraction of individual Cr fibers (Fig. 3.6). These samples were then stored in the vacuum chambers to prevent oxidation of the Cr fibers.

Figure 3.6: SEM micrograph showing Cr fibers as obtained after chemically etching the matrix

3.4.2 In-situ test set-up

The test setup for performing in-situ small-scale mechanical tests is described in detail in the works of (Gianola & Eberl, 2009; Sedlmayr, 2012). The test setup is similar in the concept to the one used by Rzepiejewska-Malyska (Rzepiejewska-Malyska et al., 2008). It employs a FIB/SEM dual beam system which is equipped with a nanomechanical force sensor (Typ FT-S1000, FemtoTools AG, Buchs/ZH, Switzerland), X, Y and Z nano-positioning stages (Attocube Systems AG, Munich, Germany), a 3-axis nanomanipulator (Kleindiek Nanotechnik GmbH, Reutlingen, Germany) to move, align and manipulate the fibers. An overview of test set-up is shown in Fig. 3.7 along with higher magnification SEM micrograph of the force sensor Si tip and the TEM grid.

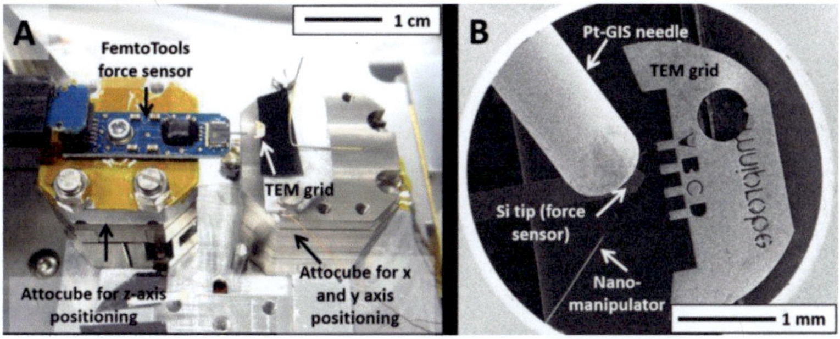

Figure 3.7: (a) Overview of the in-situ test setup indicating important components and (b) higher magnification SEM micrograph of the TEM grid and Si tip of the force sensor along with nano-manipulator and Pt-GIS needle (courtesy Charlotte Ensslen)

The dual beam SEM/FIB system is already described in section 3.2.2. The maximum force achievable with this transducer is 1000 µN and has the sensitivity of 500 µN/V. The three axis nano-manipulator is used to select and transfer the fibers from the specimen to the setup and place the fibers on the setup. The gas injection system available on the dual beam SEM/FIB system, was used to deposit platinum stripes to fix the ends of the fiber on the setup in the absence of any mechanical gripping in the system.

3.4.3 Test procedure

The first step for the in-situ test is the selection of a suitable fiber, which is long enough for the test, has a uniform diameter and can be easily manipulated with the nano-manipulator. After a suitable fiber is identified, the 3 axis nano-manipulator is carefully maneuvered to bring it close to the fiber. Pt deposition is then used to fix one end of the fiber on the nano-manipulator. The fiber is then cut free from the other end by FIB milling and the nano-manipulator is moved to the setup where the fiber is fixed on the setup using combination of FIB cuts and Pt deposition. Very small beam currents were used for cutting the fibers and it was always ensured that the area of the specimen exposed to the beam was never included in the gauge section of the test specimen. The process of harvesting, manipulation, transfer, gripping and testing of the Cr fibers is depicted in the Fig. 3.8. The setup of the Cr fiber for in-situ test is shown in Fig. 3.9.

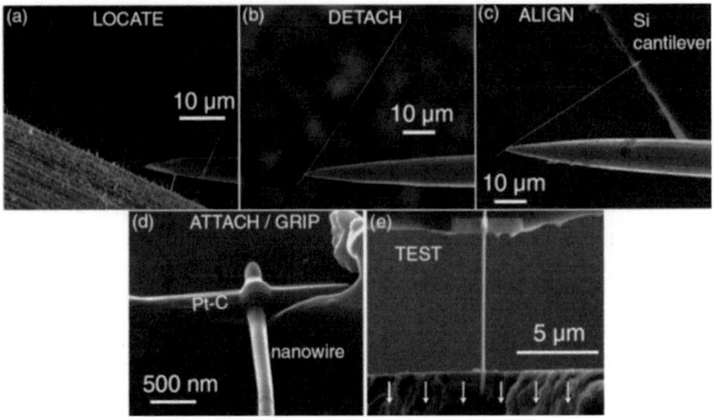

Figure 3.8: SEM images showing different steps involved in the harvesting, manipulation and testing of a nanowire/fiber in a typical in-situ tensile test (Gianola et al., 2011)

Tensile test of the fibers were conducted at constant displacement rates by increasing the piezo-voltage in the actuator until fracture of the fiber while tracking the force signal simultaneously. All tensile tests were performed with a nominal strain rate of $1x10^{-3}$/s. After fracture of a fiber, if one of the broken segments of the tested Cr fiber was found to be long enough, it was re-gripped on the set-up using the same method and tested again. Furthermore, fibers of different diameters from samples prepared at different solidification speeds were tested to examine the effect of the diameter on the strength and the deformation behaviour of the Cr fibers. During the test, SEM images were taken in regular intervals and used for strain calculation by digital image correlation (DIC).

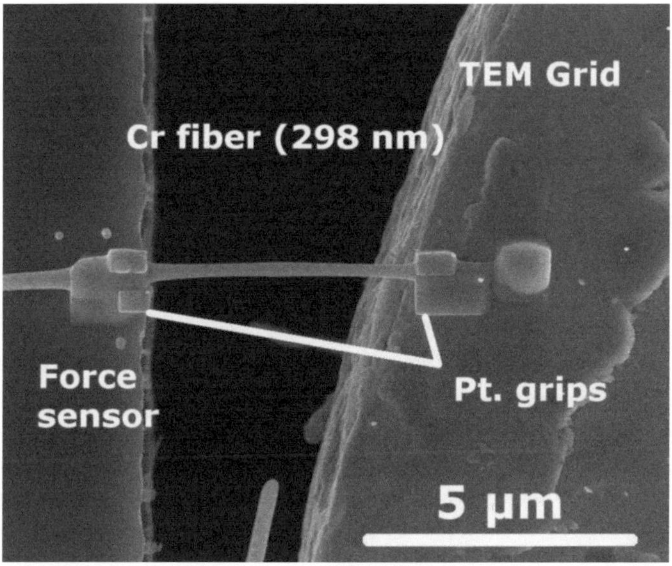

Figure 3.9: SEM micrograph of the test set up for in-situ SEM tensile test showing a Cr fiber attached to a TEM grid and a force sensor by Pt. deposited grips

3.4.4 Strain calculation by DIC

Digital image correlation has become a standard non-contact method for micro- and nano-scale testing where a direct and accurate measurement of local displacement or strain is not possible due to small size of the setup and the high resolution required. This technique which is commonly applied to macroscale testing using white light microscopy for imaging, has been extended to micro- and nano- scale testing where images are obtained by SEM or other techniques (Biery, de Graef, & Pollock, 2003; Richter et al., 2009; Sutton, Li, Garcia, et al., 2007; Sutton, Li, Joy, Reynolds, & Li, 2007).

In this work a custom software suite, which uses MATLAB® as a calculation engine, developed by C. Eberl, D. Gianola and R. Thompson (Eberl,

Gianola, & Thompson, 2006) at Johns Hopkins University, Baltimore is employed for strain calculation. Natural contrast present in the Cr fibers was employed in these tests to track the deformation of the Cr fibers. The method is described in detail in the work by M. A. Sutton and co-workers (Sutton, Li, Garcia, et al., 2007; Sutton, Li, Joy, et al., 2007).

DIC is a technique which relies on the intensity variation in the digital images, which can be natural or artificially introduced on the specimen surface, for the strain calculation. DIC works, in general, by calculating the maximum correlation between the consecutive images or subset of images which are shifted relative to each other. An iterative approach is employed to maximize the correlation coefficient by using a correlation algorithm provided by MATLAB®.

Due to the small size of the fibers, general methods to introduce surface contrast to the fiber could not be applied as they are more suited for larger specimens. Single crystalline Cr fibers also do not provide enough contrast for accurate strain measurement using DIC. Therefore, the deposited Pt stripes at the ends of the fiber, used to grip the fiber, provided better contrast and was used in this work. Markers were placed on the Pt grips on both sides as shown in Fig. 3.10. The strain in the samples was evaluated by using the '1D average strain' tool in the program, where the horizontal displacement, obtained by subtracting the current position of the markers from the initial position of the markers, is plotted versus the current position. The '1D average strain' tool gives the strain-image data. Using the 'strain-image' data and 'stress-time' and 'time-image' data set acquired during the test, 'stress vs strain' data is obtained and plotted.

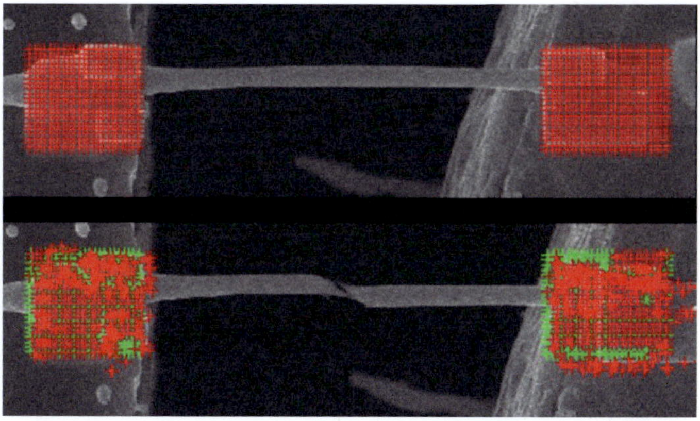

Figure 3.10: SEM micrographs of a Cr fiber (298 nm in diameter) before (a) and after (b) the test showing the grid on the Pt grips used to calculate the strain by DIC.

3.5 Micro-pillar compression test

As a major goal of this study is to characterize individual components of the eutectic composite. For this, micro-pillars were prepared by FIB milling in a dual beam SEM/FIB and then tested in a nanoindenter using a flat punch. Three types of pillars were prepared for this purpose:

Matrix only NiAl pillars (M.O.): Single phase NiAl matrix only pillars were prepared by selecting the matrix volume between the Cr fibers. In order to prepare a pure single phase NiAl pillar without any contribution from the fibers, the size of the pillars were limited by the spacing between the fibers. Furthermore, since the fiber spacing decreases with increasing solidification speed of the samples during directional solidification, the size of these pillars were also dependent on the samples from which they were prepared. Therefore, the top diameters for ORNL_20 samples were ~ 600 nm, while for the faster solidification speed samples, the diameters for the ORNL_50 and ORNL_80 samples were smaller of ~400 nm and ~300 nm respectively.

Single fiber composite (S.F.) pillars: Pillars containing a single fiber surrounded by the matrix was prepared to represent the composite at small scale and to test the interface. Similar to the M.O. pillars, the diameter of the pillars prepared for testing were limited by the fiber diameter and spacing between the fibers, which is dependent upon the solidification speed of the samples. Thus, the pillar diameter decreases with the samples for higher solidification speeds. The average top diameter of these pillars was 1.02 µm, 0.83 µm and 0.66 µm for ORNL_20, ORNL_50 and ORNL_80 samples, respectively. But due to inclusion of the fiber as well as the matrix surrounding the fiber, the diameters of these pillars were about twice as for the M.O. pillars for the same solidification speed.

Multiple fiber composite (M.F.) pillar: Since for the single phase pillars and S.F. pillars, the diameter of the pillars depend upon the solidification of the samples and, therefore the pillars from different samples have different diameters and are not ideal for comparison among varying solidification speeds as size effect on the strength of the small scale samples becomes important (Oliver Kraft et al., 2010; Uchic & Dimiduk, 2005). To remove the above mentioned size effect and to be able to compare the strength of micro pillars from different solidification speed samples, pillars of 4 µm diameter were prepared on samples from all solidification speeds, containing multiple fibers in a pillar.

3.5.1 Micro-pillar fabrication

The process used for micro pillar fabrication is a two-step process. First, a coarse structure is prepared, with a diameter of 35 µm leaving an island with a diameter of a few micrometers larger than the desired pillar sizes as shown in the Fig. 3.11. The coarse structure is prepared at high ion beam current to reduce the time required to mill the coarse structure since the aim is remove large volumes of material in the shortest possible time. The accuracy of this coarse cut is not a concern. Nevertheless, one should be careful using high ion beam currents since at high currents there is a considerable increase in the spot size and lateral width of the beam, and the beam should be focused carefully.

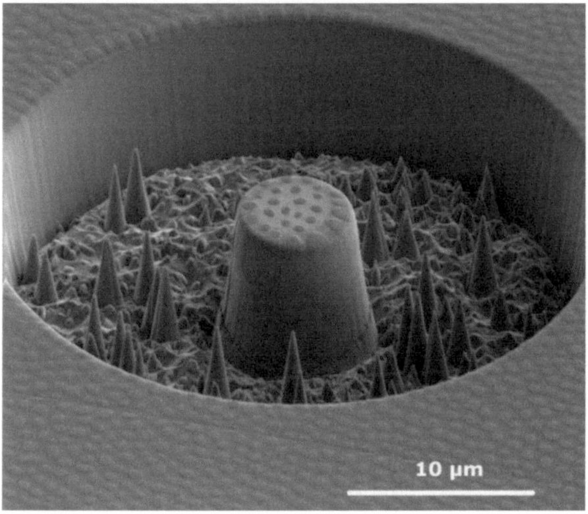

Figure 3.11: SEM image of the course structure for micro-pillar preparation, milled at high beam current to remove the excess material fast while leaving an island for fine milling with low beam current

The coarse structure has following distinct advantages:

1. It helps for locating the pillars in the nanoindenter system, where the magnification of the microscope is limited.
2. It allows for imaging of the pillars during formation or after the tests.
3. It ensures that the tip of the nanoindenter would not touch any other point of the sample and, thus, ensuring that the full load from the tip is experienced by the pillar.
4. It reduces the chances of redepositing the sputtered material back onto the surface of the pillar.

In this work, there was added difficulty to the milling because of the difference in the milling rates for the fiber and the matrix. Therefore, the milling parameters had to be adjusted to compensate for that. It was found that a beam current of 5nA along with a higher beam dwell time produced good results, while going for higher beam currents gave unsatisfactory results. The coarse structures for the pillars were produced in batches of 16 in a single FIB session, usually overnight. For the M.O. pillars and S.F. pillars, the island of the course structure was 4 μm in diameter while for the M.F. pillars, the coarse structure was 10 μm in diameter.

Once the coarse structures were milled, the finer milling of the pillars into the desired size was done using finer spot sized low beam currents. The annular milling process was used to prepare the micro-pillars used in this study, the details of which has been explained in section 2.5. For the M.O. pillars and S.F. pillars, a beam current of 0.1 nA with a dwell time of 2 ms was found to mill to the depth of about 2 μm in the material. Since the size of the NiAl pillars and S.F. pillars were in the range of 600 nm to about 1.2 micrometer, a depth of about 2 μm was enough to achieve the desired pillar height to diameter ratio between 2:1 and 3:1. For the larger pillars in the M.F. pillars, 2 passes of 0.5 nA current had to be used to achieve the required depth to be in the desired pillar height-to-diameter ratio. The final prepared pillars are shown in Fig. 3.12.

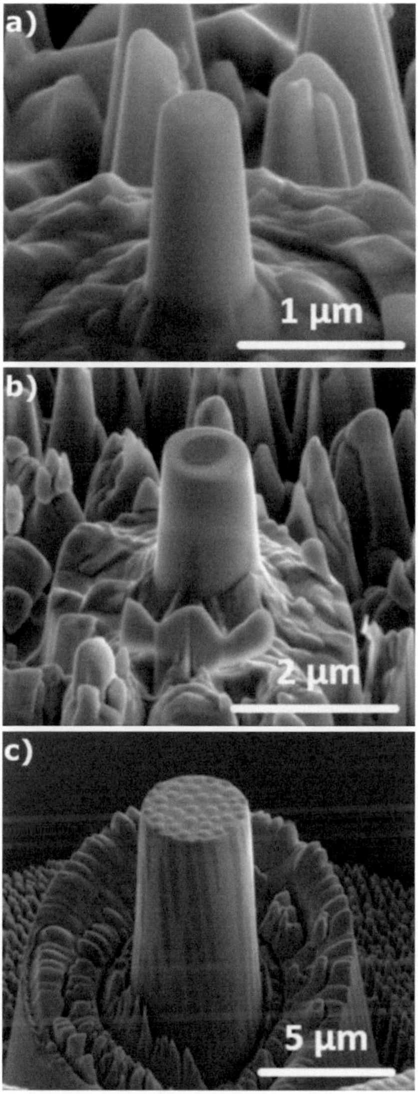

Figure 3.12: SEM images of the final prepared micro-pillars on ORNL_20 samples, a) M.O. pillar consisting of NiAl matrix, b) S.F. pillars consisting of a single fiber surrounded by the matrix, and c) M.F. pillars with a diameter of 4 µm containing several fibers.

3.5.2 Compression of micro-pillars

The micro pillars prepared by FIB milling as described above were compressed in a nanoindenter G200 (Keysight Technologies GmbH, Böblingen, Germany) using a flat punch of 10 µm diameter. The pillar compression tests were carried out in displacement control, by using a feedback loop from the force signal, since the G200 is inherently a load controlled machine. Using the special mounting blocks mentioned in section 3.3, which allows the sample on an SEM holder to be transferred to the nanoindenter, the sample containing the micro-pillar batches were loaded on the nanoindenter sample stage. The maximum indenter displacement of 25% of the pillar height was selected in order to ensure that all the pillars were compressed beyond their yield stress and were plastically deformed. The displacement rate of the indenter head was adjusted for each batch of the pillars according to the height of the pillars so that the strain rate in the pillars corresponded to about 1 x 10^{-3} /s according to the relation given by:

$$strain\ rate\colon \dot{\epsilon} = \frac{v}{h} \tag{3.6}$$

where,

v = displacement rate of the indenter head

h = Height of the pillar

The test starts with the slow approach of the indenter head towards the pillar until it encounters the top surface of the pillar, which is identified by the sudden change in the measured stiffness. At this point the indenter pushes on the pillars with the prescribed displacement rate of the indenter head until the maximum defined displacement is reached. At this maximum displacement, the maximum load is held constant for 10 seconds, and then the pillar is unloaded at a rate of 100 µN/s

After the compression test of the pillars, all samples were examined post-mortem using the dual beam SEM/FIB system to characterize the deformed pillars. A few of the deformed pillars from the S.F. pillars group were selected and cross cuts through the pillars were prepared to examine the deformed interface. Furthermore, a few of the deformed pillars were selected to prepare lamellae to be examined by TEM to characterize the active slip systems and to understand the dislocation evolution during the deformation.

3.5.3 Analysis procedure

The load-displacement data from the nanoindenter system obtained for the compression testing of the pillars were exported to Origin® software for the analysis. There are various ways in which the micro pillar compression tests can be analysed and a comprehensive comparison of different methods can be found in (Fei, Abraham, Chawla, & Jiang, 2012). It is important to use a consistent approach for the analyses of the micro pillars in order to compare different samples. For this study, the load on the pillar was converted to stress using the top diameter of the pillars:

$$\sigma = \frac{P}{\pi r_T^2} \tag{3.7}$$

with,

σ = stress in the pillar,

P = load on the pillar, and

r_T = radius of top of the pillar.

The displacement data obtained from the nanoindentation system has contributions from the deformation of the indenter, the pillar and the substrate. Therefore, the directly measurable 'total displacement' Δx_{total} consists of:

$$\Delta x_{total} = \Delta x_{pillar} + \Delta x_{indenter} + \Delta x_{substrate} \quad (3.8)$$

Thus:

$$\Delta x_{pillar} = \Delta x_{total} - (\Delta x_{indenter} + \Delta x_{substrate}) \quad (3.9)$$

Therefore, according to (Fei et al., 2012) the total displacement of the pillar can be calculated as:

$$\Delta x_{pillar} = \Delta x_{total} - (\frac{(1-\upsilon_{ind}^2)P}{2*E_{ind}*r_{top}} + \frac{(1-\upsilon_{sub}^2)P}{2*E_{sub}*r_{base}}) \quad (3.10)$$

where,

$\upsilon_{ind,\,sub}$ = poisson's ration of indenter and substrate respectively

$E_{ind,\,sub}$ = Young's Modulus of indenter and substrate respectively

r_{top} = radius of the top of the pillar

r_{base} = radius of the base of the pillar

The strain in the pillar can then be calculated by dividing this displacement by the initial height of the pillar:

$$\epsilon = \frac{\Delta x}{h_{pillar}} \quad (3.11)$$

A representative stress-strain curve for one of the S.F. pillars in ORNL_20 sample is presented in Fig. 3.13.

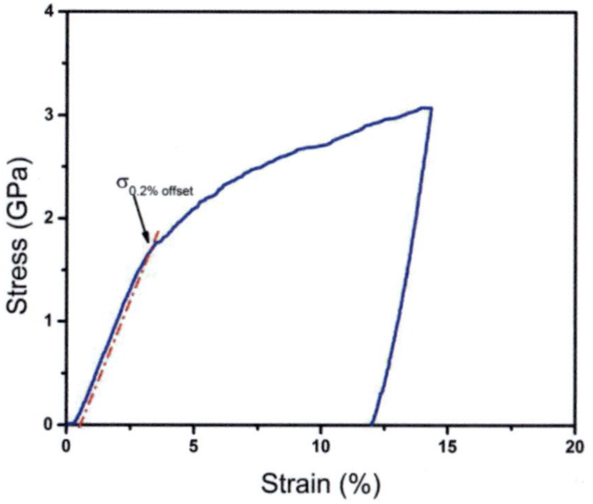

Figure 3.13: Representative engineering stress vs strain response obtained from a micro-pillar compression test of a S.F. pillar on ORNL_20 sample

The stress strain curve shows an elastic loading segment, followed by plastic deformation at about 2 GPa when the slope of the curve is changing. Since there was no sharp transition, which could be identified as the onset of plastic deformation, a 0.2% offset yield strength method was applied to all curves to estimate the yield strength $\sigma_{0.2\%}$ of the pillars. A straight line parallel to the largest slope of the loading curve in the elastic segment was drawn. This line was then shifted by 0.2 % strain and the intersection to the stress-strain curve was taken as the yield strength of the pillar. The yield strength of all pillars of same type and size were then averaged for that particular sample.

3.6 TEM examination

With the help of A. Krüger (IAM-AWP), a participating institute of the IMD graduate school, transmission electron microscopy (TEM) examinations of both the undeformed and deformed NiAl-Cr samples were performed. The undeformed samples were prepared by standard TEM preparation methods, dimple grinding followed by ion polishing. The TEM lamellae of the deformed composite pillars were prepared by FIB milling (Dual Beam FIB/SEM Workstation, Nova NanoLab 200, FEI Company, Hillsboro, OR, USA) following standard procedures. The lamellae from deformed pillars were prepared by D. Exner (IAM-WBM). The lamellae were prepared from S.F. pillars in order to observe the dislocation structure resulting from the deformation process. The TEM examinations were carried out with a Tecnai F20 (FEI Company, Hillsboro, Oregon, USA) operated at 200 KeV acceleration voltage and CM30 (FEI Company, Hillsboro, Oregon, USA) operated at 300 KeV.

4 Results and discussion

4.1 Microstructural analysis

The SEM examination in the transverse and longitudinal direction on the ORNL_20, ORNL_50, and ORNL_80 revealed a fiber-matrix microstructure with NiAl forming the matrix and Cr is in the form of long continuous fibers. SEM images from all the samples in transverse and longitudinal directional are shown in Fig. 4.1. The fiber diameter and fiber spacing decreases with the increase in the solidification speed similar to that reported in the literature (Rablbauer et al., 2004; Walter & Cline, 1970). An image manipulation software, ImageJ®, was used to calculate the fiber diameter and fiber spacing, from the SEM micrographs of the transverse section for all the samples. The fiber diameter for ORNL_20 is around 690 nm whereas for ORNL_80, it goes down to around 310 nm. The spacing between the fibers for ORNL_20 is around 1.2 µm which reduces to around 500 nm for ORNL_80 samples (table 4.1).

Examinations in the longitudinal section confirms that Cr forms long continuous fibers. The fibers are aligned parallel to the growth direction. TEM examination by A. Krüger (IMD, RA V) confirms the growth direction for both matrix and fiber is [100]. Also prominent is the existence of cellular structures with increase in the solidification speed. The alignment of the fibers at these cell boundaries is disturbed and the fibers are tilted towards the boundary as shown in fig. 4.2

Table 4.1: Average fiber diameter and average fiber spacing for the three solidification speed samples. Image manipulation software ImageJ® was used to analyse the SEM micrographs of the transverse section of the samples to obtain the values

Solidification speed (mm/h)	Fiber diameter (μm)	Fiber spacing (μm)
20	0.68±0.08	1.12±0.16
50	0.51±0.08	0.70±0.03
80	0.31±0.04	0.48±0.19

4.2 Nanoindentation

Nanoindentation study on the ORNL samples revealed that the elastic modulus of the three samples was constant and the average value was found to be 210-215 GPa. The elastic modulus values reported in the literature (G. Frommeyer, Rablbauer, & Schäfer, 2010) for the same composition matches very well with the results from the study. The hardness values for the three alloys were also found to be constant for the different solidification speeds and were in the range of 4.5 GPa. The values of hardness and elastic modulus are tabulated in table 4.2 and also compared in Fig. 4.3.

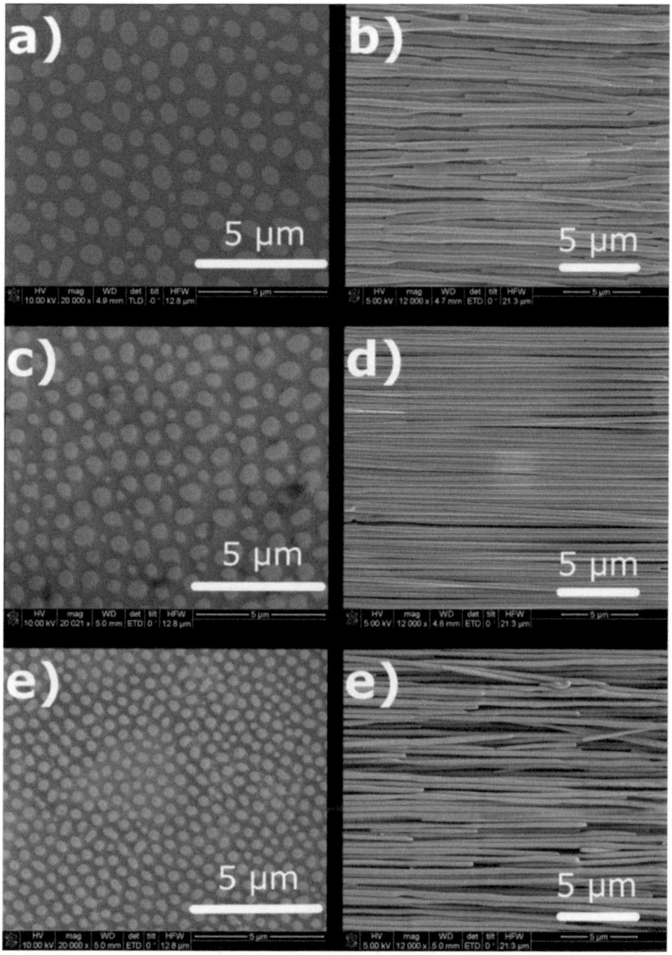

Figure 4.1: SEM images of NiAl-Cr DS eutectic alloy in transverse direction showing matrix with dark contrast and fibers with light contrast, and longitudinal direction after matrix etching showing continuous fiber aligned in the growth direction for ORNL_20 (a, b), ORNL_50 (c, d), and ORNL_80 (e, f) respectively

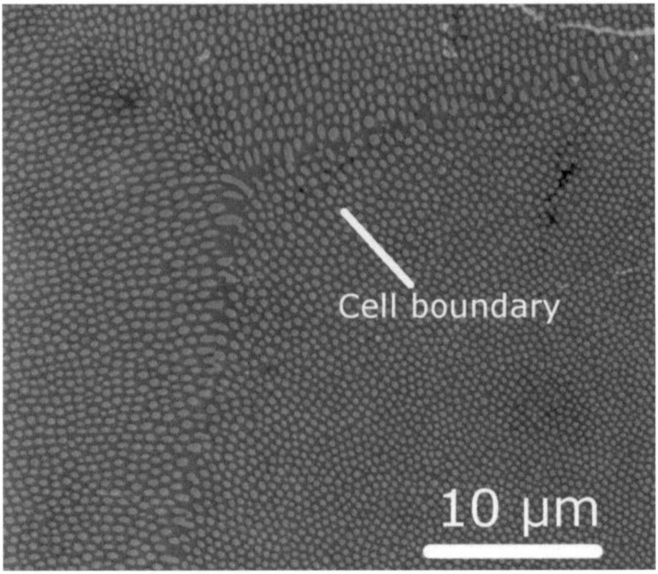

Figure 4.2: SEM micrograph of transverse section of ORNL_80 samples showing the cell boundary where the alignment of the fibers s disturbed

Table 4.2: Hardness and Young's modulus of the three solidification speed samples. The values are based on about 20 individual indents and analysed as described in section 3.3. The errors correspond to the standard deviation

Solidification speed (mm/h)	Hardness (GPa)	Young's modulus (GPa)
20 mm/h	4.59±0.01	213.8±3.0
50 mm/h	4.45±0.09	221.6±8.3
80 mm/h	4.65±0.16	215.3±5.4

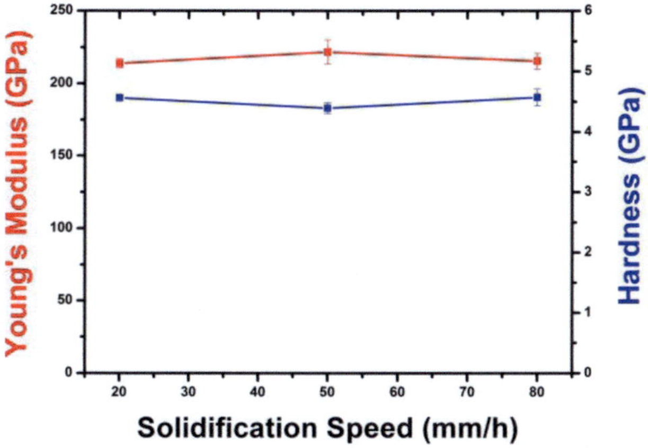

Figure 4.3: Young's modulus and hardness vs solidification speed of the three samples, showing hardness and modulus being independent of the solidification speed

The constant value of elastic modulus with respect to the solidification speed is expected and can be simply explained due to the same volume fraction of the NiAl and Cr in all the three samples as the composition is same for all the three alloys. But the constant values for hardness in all the three samples was not expected as the fiber diameter and fiber spacing is changing with the solidification speed. This unexpected result highlights the need to develop a mechanistic understanding of the deformation mechanisms in these alloys and motivates to characterize the different phases and interfaces between the two phases in order to fully understand their role for the deformation mechanisms.

4.3 SEM in-situ tensile test

Individual Cr fiber, after being extracted from the sample by selective chemical etching, were transferred and tested in tension as described in section 3.4. The initial diameter of the fiber was used to estimate the stress in the fibers and the strain measurement was done by DIC. A typical stress-strain curve for a fiber of diameter 298 nm, is shown in Fig. 4.4a along with the SEM micrographs from different stages of the test (Fig. 4.4 b-e). The fiber undergoes elastic elongation upon application of load (Fig. 4.4 b). At the peak load the SEM micrographs reveal that a layer was formed surrounding the fiber which fails in a brittle manner (Fig. 4.4 c). Until this layer failed it was not possible to observe any deformation in the fibers during the in-situ tests in SEM. Soon after this outer layer failed, local necking in the Cr fiber occurs (Fig. 4.4 d) followed by final separation of the fiber (Fig. 4.4 e).

TEM examination of one of the failed Cr fibers was performed to characterize the outer layer that was formed surrounding the fiber. It was revealed that this outer layer was mainly carbon (Fig. 4.5). It is believed that, although it is not possible to observe any deformation in the Cr fiber before fracture of the outer layer in SEM during the in-situ test, it is the local plastic deformation in the Cr fiber which leads to the failure of the outer layer, rather than the other way around. Also the load bearing capacity of the Cr fibers would not be much affected by this outer layer of carbon, since carbon, deposited inside an SEM, is much weaker than the Cr (Zhang et al., 2010). Nevertheless, due to this outer layer, the diameter of the Cr fiber measured from the SEM images may be overestimated, which will lead to an underestimation of the stress levels observed in the fibers. Therefore, it is safe to say that the stress levels observed in the Cr fibers reflects a lower bound value. Although it is very difficult to measure the thickness of this outer layer, a rough estimate leads to the layer thickness of ~15 nm, for a fibers of measured diameter 298 nm. Using the approach suggested by Gianola et al. (Gianola et al., 2011), the error in measurement can be estimated up to ~15 %.

All the Cr fibers showed high strength values in the in-situ tensile tests in the range of 1-3 GPa. The ultimate tensile strength of the Cr fibers is plotted against the fiber diameter in Fig. 4.6. Despite these high values for tensile strength observed in the Cr fibers, strengths close to theoretical strengths as overserved for the case of Mo fibers (Johanns et al., 2012) were not observed. The near-theoretical strength in Mo fibers were only observed for the dislocation free Mo fibers, while pre-strained fibers having higher dislocation densities showed lower strengths. The strength of the Cr fiber were also found to increase with decreasing fiber diameter. This size effect on the strength of Cr fibers having sub-micron dimensions can be attributed to truncation or lack of dislocation sources with decreasing specimen size and as introduced in section 2.4. The high stress levels observed in these Cr fibers as well their dependence on the diameter of the fibers suggests that the un-deformed Cr fibers have a very low initial dislocation density as suggested for Mo fibers (Phani et al., 2011) and observed for Al fibers (Mompiou et al., 2012). The plastic deformation in these fibers start at the site having the most favourable dislocation configuration and, thus, would obey weakest-link statistics and may be described by a Weibull distribution which has also been discussed in detail in the work of Senger et al. (Senger et al., 2011). Furthermore, it can also be observed in Fig. 4.6 that the strength of the re-attached fibers were found to be higher than the pristine fibers. This observation also supports the weakest-link argument since in the re-attached fibers, the next weakest site will determine the strength.

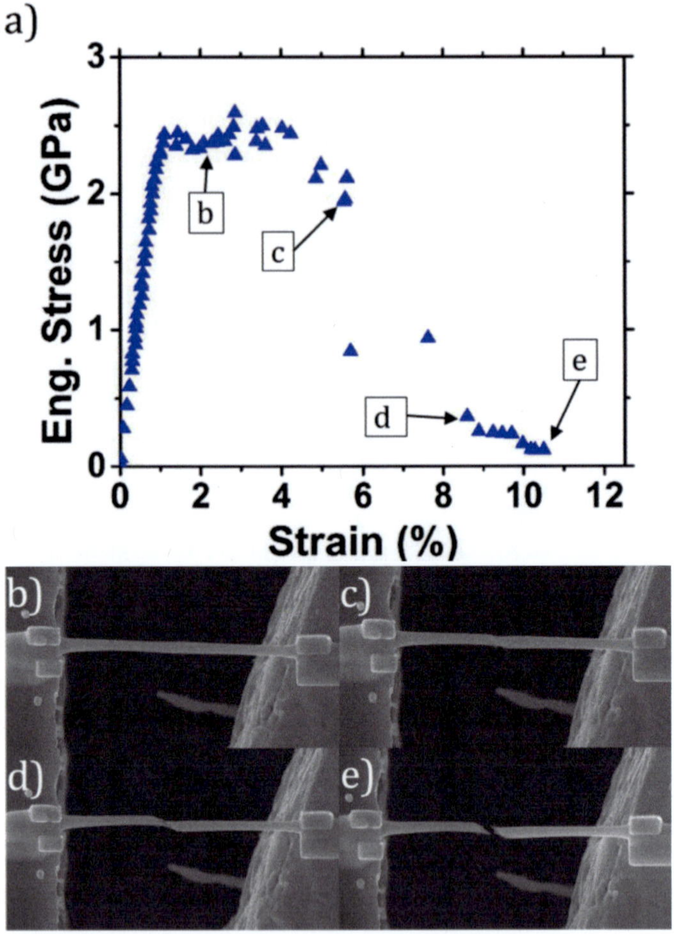

Figure 4.4: a) Engineering stress vs strain curve of an individual Cr fiber (298 nm) obtained from in-situ SEM tensile test, b) – e) SEM micrographs from different stages of the test marked on the stress-strain curve

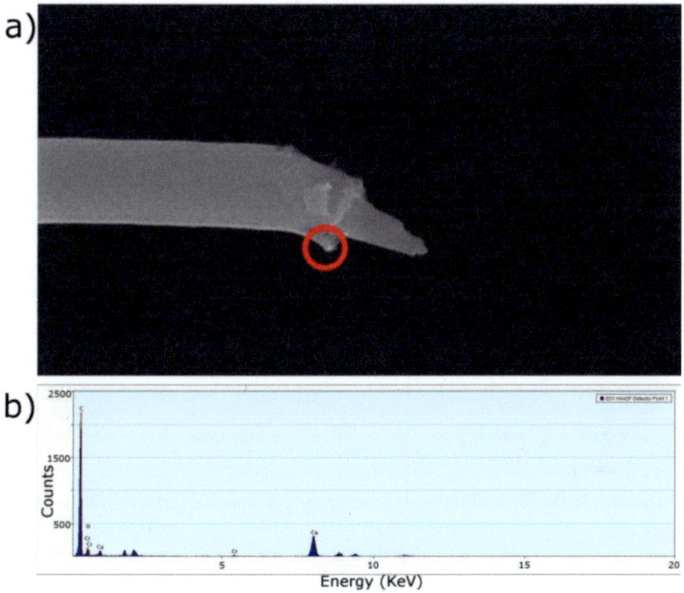

Figure 4.5: a) SEM image of the failed Cr fiber showing a fractured outer layer. The chemical composition was examined at the site marked by the red circle and b) results of a TEM EDX analysis confirming that the outer layer is carbon

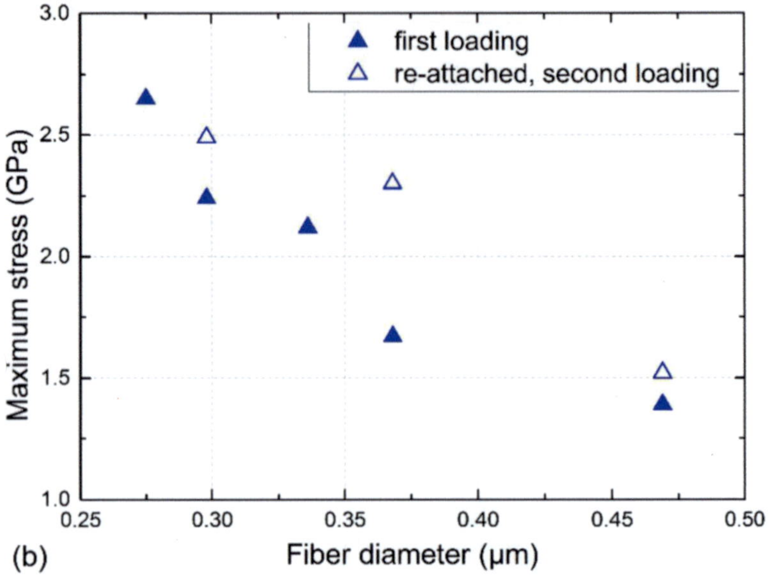

Figure 4.6: Plot of maximum stress vs fiber diameter showing the increase in the strength levels with decreasing diameter. Re-attached fibers show higher strength than the pristine fibers.

4.4 Micro-pillar compression tests

4.4.1 Stress-Stain response

Typical stress-strain responses for the three pillar types from ORNL_20 are shown in Fig. 4.7. All three pillars show high strength values with of M.O. pillar showing highest yield strength whereas the M.F. pillars show lowest strength values. After the yield both M.O. and S.F. pillars show hardening behaviour while it is not seen in the M.F. pillars. An instability is observed in the M.F. pillars following yield, which is believed to be a result of minor misalignment between the pillar and the flat punch. The

0.2 % offset yield stress and the average diameter of the pillars with respect to the different sample are tabulated in table 4.3 and plotted in Fig. 4.8. All pillars show high strength values up to 4 GPa. It can be noted that for the same type of pillars (M.O., S.F. or M.F. pillars) the strength of the pillars increases with the increasing solidification speed of the sample. The increase in strength of the pillars with solidification speed in the S.F. pillars and the M.O. pillars is attributed to the decrease in the diameter of the pillars. Obviously, the pillars needed to be prepared with smaller diameters with increasing solidification speeds due to the reduced spacing between the fibers. Therefore, a size effect due to dimensional constraint occurs, as already discussed in section 2.4. The 0.2 % offset yield stress of all the M.O. pillars and S.F. pillars were plotted against their diameter and the two pillar groups were fitted with power law fit as shown in Fig. 4.9 to evaluate the scaling behaviour of these pillars. It was observed that the scaling behaviour of the fitted curve for S.F. pillars (b= -0.39) show weaker trend than for the M.O. pillars (b= -0.57). If we compare the strength of the two pillar groups having diameter around 600 nm, it can be seen that the M.O. pillars are somewhat stronger than the S.F. pillars.

The M.F. pillars show an increase in the 0.2 % offset yield strength with increasing solidification speed. The average 0.2 % offset yield strength of the M.F. pillars for the three samples are also shown Fig. 4.9. There is a weak trend that the strength increases with increasing solidification speed. It can also be seen that the strengths of the M.F. pillars are lower than the strengths of the S.F. pillars and M.O. pillars prepared from the same sample, as expected due to the increase in size.

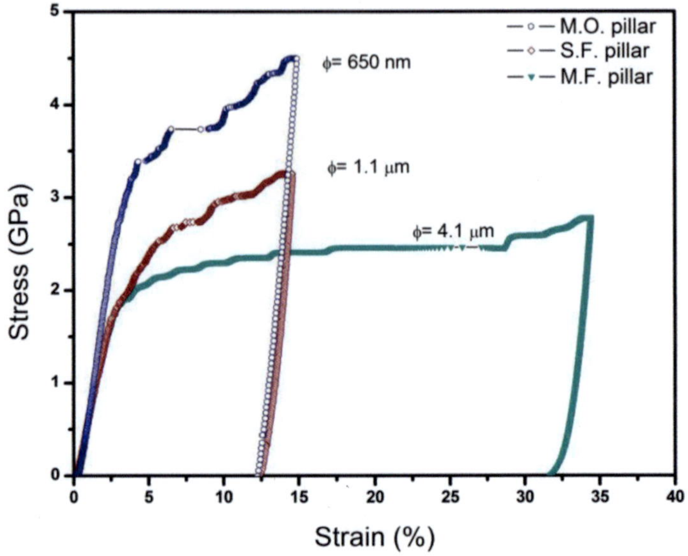

Figure 4.7: Representative stress vs strain plot for the three different types of pillars for ORNL_20 sample. M.O. pillars show highest strength values while M.F. pillars show instability after yield. The diameters of the pillar are also indicated along with the plots.

Table 4.3: Average 0.2% offset yield stress and the average diameter of the three pillar types with respect to different samples, at least 5 μ-pillars for each condition were tested and the error bars correspond to the standard deviation.

Solidification speed	**Matrix-only pillars**	**Single-fiber pillars**	**Multi-fiber pillars**
	Diameters (μm)		
20 mm/h	0.65±0.08	1.02±0.10	4.07±0.04
50 mm/h	0.52±0.07	0.83±0.05	4.05±0.02
80 mm/h	0.39±0.03	0.66±0.03	4.09±0.02
	0.2% offset yield stress (GPa)		
20 mm/h	2.51±0.28	1.77±0.22	1.61±0.05
50 mm/h	2.74±0.52	1.98±0.41	1.71±0.01
80 mm/h	3.33±0.36	2.29±0.42	1.94±0.30

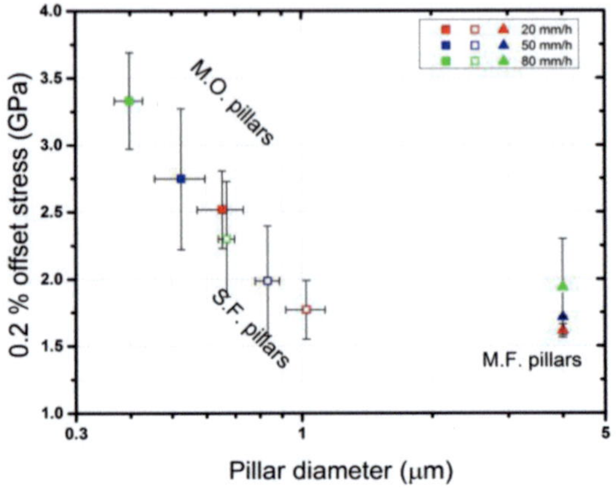

Figure 4.8: 0.2% offset stress of the three pillar groups for the different samples vs the pillar diameter

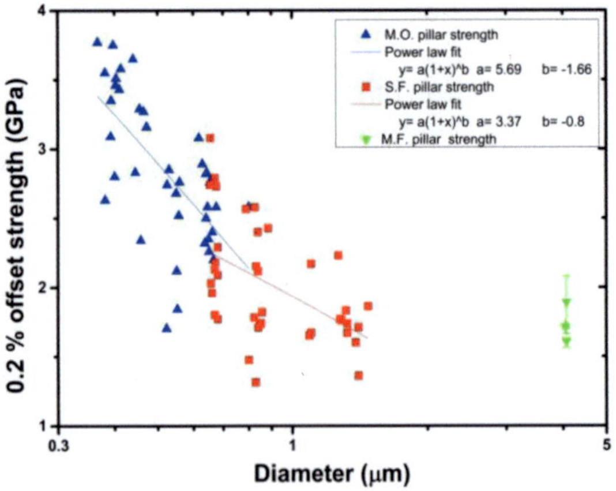

Figure 4.9: 0.2% offset yield stress for all the pillars tested vs diameter of the pillars. The M.O. pillars and S.F. pillars were fitted with a two-parameter power law fit which shows that the S.F. pillars show a weaker trend than the M.O pillars. The yield strength of M.F. pillars is also shown in the plot

4.4.2 Post-mortem examination

After the compression tests, all three types of pillars from the different samples were examined by dual beam SEM/FIB to check for any issues with misalignment or instability. Examples of deformed pillars for the three types of pillars from ORNL_20 samples are presented in Fig. 4.10 and compared with the undeformed pillars.

The M.O. pillars and the S.F. pillars from all three samples deformed in a similar manner. A small number of non-uniformly distributed distinct slip steps were observed on the surface of the M.O. pillars (Fig. 4.10 b). In the case of the S.F. pillars, either no slip steps were observed or a large number of uniformly distributed small slip steps were seen (Fig. 4.10 d).

Dimiduk et al. (Dimiduk, Uchic, & Parthasarathy, 2005) have seen a similar effect in single crystal Ni micro pillar experiments. They observed a small number of distinct slip steps on sub-micron pillars but a large number of uniformly distributed slip steps on bigger pillars. They have explained this observation on the basis of differences in the dislocation multiplication mechanisms in the samples with different sizes. For specimens having diameters less than 10 μm, the dislocation multiplication process is limited, especially because of the dislocation leaving the free surface before multiplying. Larger specimens show bulk like behaviour with dislocation multiplication causing the plastic deformation to be carried out by multiple slip events. The difference between the M.O. pillars and S.F. pillars, although quite different in size, is much less pronounced compared to Dimiduk et al. However, essentially two different type of samples are tested, a single crystal in one case and a fiber-matrix composite in the other. This has two implications. First, in the composite S.F. pillars, a dislocation cannot easily travel across the entire pillar since it needs to overcome or bypass the fiber. Second, in the S.F. pillar the interface between the matrix and fiber may serve as a dislocation source.

The post-mortem examination of the M.F. pillars revealed that all pillars had failed by buckling (Fig. 4.10 f). This could be related to some misalignment between the indenter tip and the pillar surface. It can be also argued that the buckling relates to a buckling instability due to the large aspect ratio of the fibers testes under compression. Nevertheless, from the stress-strain curve of the M.F. pillars (Fig. 4.6) it is clear that the plastic instability occurred at higher strains beyond yielding. Therefore, the misalignment should not have a significant effect on the stress values at the onset of yielding, as it was shown in the work by Schwaiger et al. (Schwaiger, Weber, Moser, Gumbsch, & Kraft, 2011).

Cross cuts through selected deformed S.F. pillars were prepared by FIB milling. An SEM micrograph of one of such pillars is shown in Fig. 4.11, confirming the co-deformation of both the fiber and the matrix. No sign of any delamination or fracture at the fiber-matrix interface is observed despite being heavily deformed. This strongly suggests that the interface between the fiber and matrix is in-fact very strong and that both fiber and the matrix are plastically deformed.

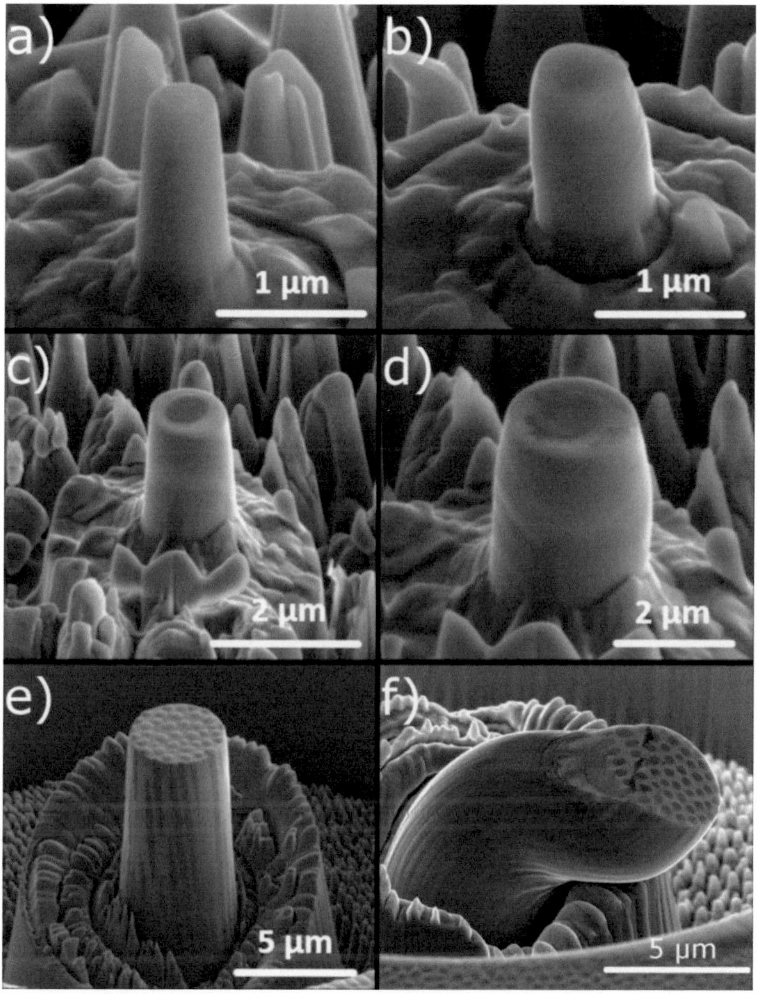

Figure 4.10: SEM images of the undeformed a) M.O., c) S.F., and e) M.F. pillars deformed b) M.O., d) S.F., and f) M.F. pillars showing distinct slip steps and non-uniform deformation in M.O. pillars, a more uniform distribution of slip in S.F. pillars and M.F. pillars failing by buckling

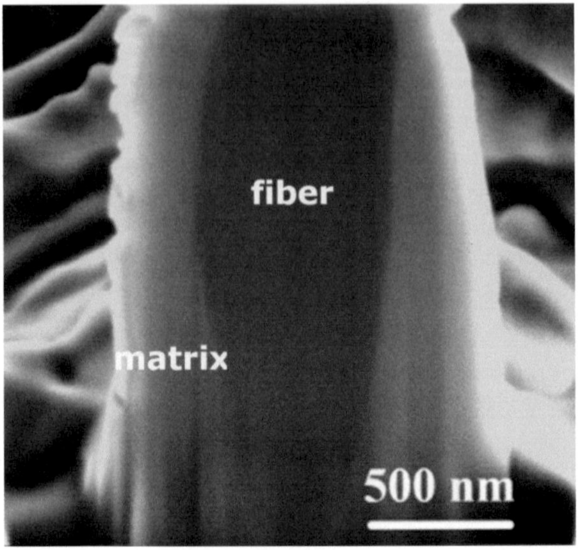

Figure 4.11: SEM micrograph of the cross cut of a deformed S.F. composite pillar showing co-deformed matrix and fiber with no sign of delamination or fracture at the interface between the fiber and the matrix, suggesting a strong interface

4.5 TEM examination

The TEM examination of the undeformed specimen in the transverse and the longitudinal section were carried out and the micrographs are shown in Fig. 4.12. The transverse section of the undeformed specimen (Fig. 4.12 a) revealed that there were few dislocations in the NiAl matrix while the fibers contained only very small number of dislocations. This supports our argument that the high strength values of the Cr fibers, observed in the in-situ tensile tests, relates to a very low initial dislocation density in the Cr fibers. Examination of the longitudinal section of the undeformed specimen reveals that the interface between the matrix and fiber is surrounded by a network of dislocations (Fig. 4.12 b). The small lattice mismatch between the NiAl and the Cr, resulting in the semi-coherent interface is responsible for the misfit dislocation network as they reduce the coherency strain produced by the mismatch across the interface. These dislocation network has also been reported in NiAl-Cr/NiAl-Cr(Mo) composites in the literature (X. F. Chen, Johnson, Noebe, & Oliver, 1995; E. Cline et al., 1971; D. R. Johnson et al., 1995b). The lattice constants for NiAl and Cr reported in the literature (E. Cline et al., 1971) are 0.2887 nm and 0.2885 nm respectively. The lattice mismatch between NiAl and Cr is calculated from the lattice constants (E. Cline et al., 1971) as 0.31%±0.02.

The lattice mismatch can also be calculated based on the dislocation spacing (E. Cline et al., 1971) from the interface dislocation network as:

$$\delta = \frac{a_{NiAl}}{d_{ID}} \tag{4.1}$$

where, d_{ID} is the average spacing between the dislocations. The dislocation spacing in the undeformed NiAl-Cr specimen was measured to be ~74 nm which results in the lattice mismatch of 0.39 % being in good agreement with the values given by Walter and Cline.

TEM examination of the lamellae prepared from the deformed composite S.F. pillar revealed increased dislocation activity both in the fiber and the matrix (Fig. 4.12 c). A Burgers vector analysis of the dislocations in these heavy deformed pillar lamellae were also carried out. Most of these dislocations could be identified as <100> type edge dislocations. Other dislocations having a mixed Burgers vector were also observed. Similar interface dislocation structures around the fiber in NiAl-Cr directionally solidified eutectics were observed previously (E. Cline et al., 1971; D. R. Johnson et al., 1995b); dislocations of <100> type were observed only in the NiAl matrix in samples deformed to low strains (< ~0.5%) and preferentially emitted from the NiAl/Cr interface.

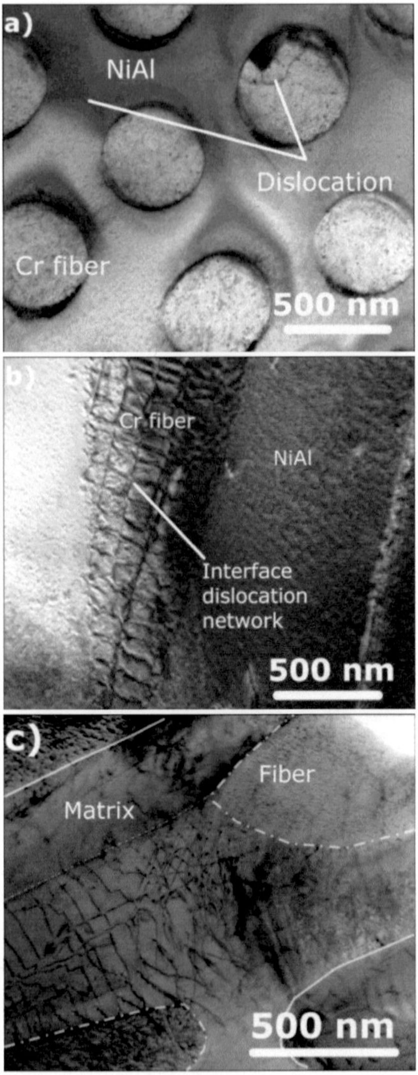

Figure 4.12: TEM micrographs of the undeformed specimens a) transverse section showing few dislocations in the matrix and fibers being almost dislocation free, b) longitudinal section with Cr fiber surrounded by network of interface dislocations, and c) section of a deformed S.F. pillar showing higher dislocation activity in the matrix as well as fibers compared to the undeformed samples.

4.6 Discussion

TEM examination of the undeformed specimen of NiAl-Cr DS eutectic reveals that the undeformed matrix and the fiber have low initial dislocation density in the fiber and the matrix. Further, the interface between the NiAl matrix and Cr fiber is surrounded by the network of a<100> type edge dislocations. The high strength of the Cr fibers and NiAl matrix observed during the in-situ SEM tensile tests and single phase pillar compression tests respectively, supplements these observations. In contrast, from TEM examination of the deformed S.F. pillars, we observe a large number of dislocations, both in the matrix and the fiber, with most of these dislocations identified as a<100> type. The strength of these S.F. pillars have been observed to be lower than the M.O. pillars of similar dimensions. The Cr fibers itself deform plastically during the tensile tests. Although, for any given sample Cr fiber, when compared to M.O. pillars and S.F. pillars, show the lowest strength among the three (Fig. 4.13), a simple rule of mixture analysis using the strength of Cr fibers and NiAl matrix pillars to calculate the strength of the composite pillars predicts higher strength values than what was observed. FIB crosscuts across the deformed S.F. pillars further reveal that the two phases, NiAl matrix and Cr fibers, are able to co-deform under stress and delamination or fracture along the interface was not observed, suggesting a very strong interface between the two phases. The lower strength of the composite pillars containing the interface along with heavy dislocation activity observed in the deformed composite pillars suggest that the introduction of Cr fibers, which are ductile in the given diameter range, in the brittle NiAl matrix enhances the room temperature ductility and toughness following ductile phase toughening concepts as discussed in the section 2.3.

A rule-of-mixture (ROM) analysis combining NiAl and Cr was performed to estimate the strength of the S.F. composite pillars from the strengths of the M.F. pillars and individual Cr fibers. The analysis was performed assuming a ratio of 0.5 for fiber-to-matrix in the cross section, since for all the pillars the ratio was in the range of 0.5±0.2. The strengths were calculated for all diameters of the Cr fibers that were tested. The procedure is as follows:

1. For each Cr fiber tested, the corresponding S.F. pillar diameter was calculated for a cross section area ratio of 0.5. Accordingly, the diameter of the equivalent M.O. pillar, having the same volume of NiAl, was calculated.
2. The experimentally obtained stress vs diameter data of the M.O. pillars and S.F. pillars tests were fitted with a power law $y = a \cdot x^b$ (Fig. 4.8).
3. From the power law fits, the strength values of the M.O. pillars were then calculated for the diameters determined in step 1.
4. Based on the strength σ_{NiAl} of M.O. pillars thus obtained and the strength σ_{Cr} of the Cr fibers, the rule-of -mixture strength σ_{ROM} of the S.F. pillars was calculated following the rule-of-mixture:

$$\sigma_{ROM} = \frac{\sigma_{Cr} \cdot A_{Cr} + \sigma_{NiAl} \cdot A_{NiAl}}{A_{comp}} \tag{4.2}$$

where A_{Cr}, A_{NiAl}, A_{comp} are the cross section areas of the fiber, NiAl, and composite pillars, respectively.

The calculated ROM composite pillar strengths were compared to the values determined for the S.F. pillars according to the power law fit (Fig. 4.14). It was found that the rule-of-mixture calculations predict a higher strength value for the composite pillars than observed experimentally. The results are summarized in Table 4.4.

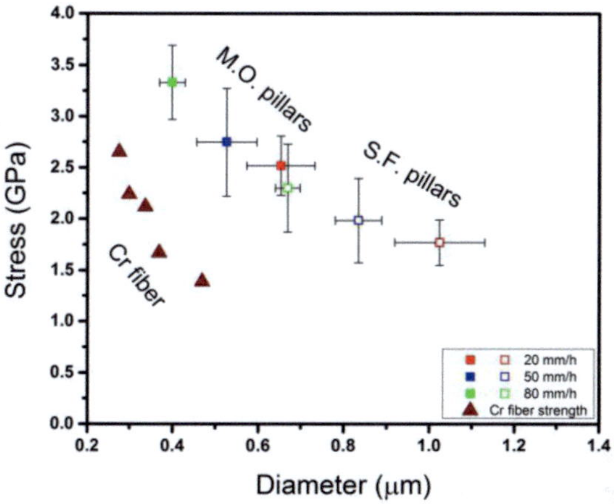

Figure 4.13: Stress vs diameter plot showing 0.2% offset yield stress for M.O. and S.F. pillars for the three samples with respect to pillar diameter. The maximum strength of Cr fiber vs fiber diameter is also shown for comparison

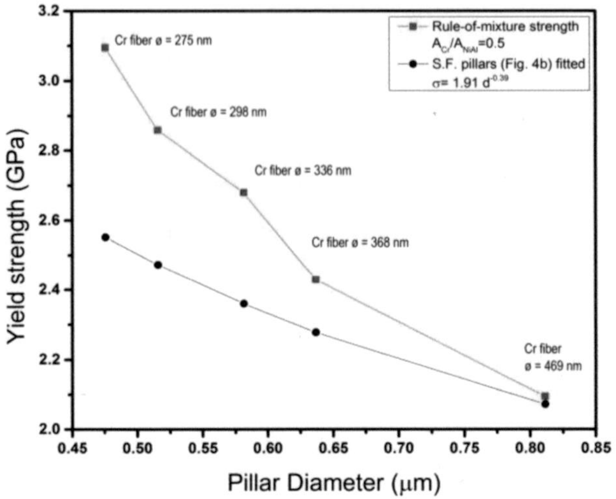

Figure 4.14: Comparison of strength of composite pillars obtained by the rule of mixture approach with the strength obtained by fitting the experimental data using a power law fit. The ROM predicts a higher strength than observed in the experiments

Table 4.4: Comparison of the strength of single-fiber composite pillars with different fiber diameters calculated following a rule-of-mixtures approach and determined from a power law fit to the experimental data

Cr fiber diameter (μm)	Strength (GPa)	
	Rule-of-mixture	Fit to experimental data
0.275	3.09	2.55
0.298	2.85	2.47
0.336	2.67	2.36
0.368	2.42	2.27
0.469	2.09	2.07

For example, a yield strength value of 3.09 GPa is predicted for a fiber diameter of 275 nm, which is higher than the yield strength value of 2.55 GPa determined from fitting the experimental data. It should also be noted that for the rule-of-mixture analysis, the compressive strength of the pillars is used with the tensile strength of Cr fibers. This may introduce two uncertainties. First, the tensile tests on longer fibers tend to show systematically lower strengths than the pillar tests (Johanns et al., 2012). Second, refractory metals show an asymmetry in strengths for tension and compression, of the order of 20% (Duesbery & Vitek, 1998). However, if an increase in compressive strength values compared to tensile strength values is considered, the predicted values for the composite pillars from the rule-of-mixture calculation would be even higher.

The difference between the strength values predicted by the ROM calculation and the experiments in figure 4.13 increases with decreasing fiber diameter. This can be attributed to the strong increase in the M.O. pillar strength for decreasing pillar diameter as seen in figure 4.8. This observation further supports the argument that the interface plays an important role in the deformation process of the DS eutectics, since the ROM calculation does not take into account the effect of the interface on the strength values of the composite pillars. It may be argued that for decreasing pillar diameters and fiber diameters, the interface area per unit volume increases and, thus, the contribution of the interface to the overall strength of the pillars increases with decreasing pillar diameter. On the other hand, for larger pillar diameters, the interface area has a smaller effect on the overall strength and it can be observed that the ROM yield strength and the fitted experimental yield strength for the largest fiber diameter (496 nm) show similar values. This is further supported by the fact that the strength of the size independent M.F pillars is comparable to the strength of the largest diameter S.F. pillars (Fig. 4.8).

To explore the mechanisms behind the ductile phase toughening, Noebe and co-workers (Noebe et al., 1991) have discussed the role of incorporating a ductile second phase into a brittle matrix for macro- and micro-scale toughening. At the macro-scale, this ductile second phase, which can be isolated particles, interpenetrating networks, or continuous lamellae or fibers, helps to increase the toughness of the brittle matrix by interfering with the propagating crack. Toughening can occur by different mechanisms, such as crack bridging, crack deflection or crack blunting.

At the micro-scale, toughening can be attributed to constrained deformation of the ductile second phase, which may induce the generation of dislocations at the interface and this enabling the deformation of brittle matrix by either direct transmission of dislocations or nucleation of dislocation in the adjacent phase. In a two phase system, the strain transfer is most likely to takes place by nucleation of dislocation in the adjacent phase. The plastic strain transfer across the interface requires an appropriate orientation relationship between the two phases and a strong interface. The constrained deformation of the ductile phase in the brittle matrix, fulfilling the above mentioned requirements would then lead to the dislocation generation at the interface. These additional dislocation generated in the interface region will then contribute to the plastic deformation of the brittle matrix and, thereby, increasing the toughness of the composite by postponing or delaying the propagation of the fracture.

The observations from different tests carried out in this study certainly confirms that the criteria described by Noebe and co-workers for microstructural toughening for a ductile second phase in a brittle matrix is fulfilled for NiAl-Cr directionally solidified eutectic alloys studied in this project. The above argument strongly suggests that it is the interface in the NiAl-Cr directionally solidified eutectics that is controlling the deformation process in these alloys. The lower strength of S.F. pillars, containing the interface region, as compared to M.O. pillars without the interface, having similar dimensions as the composite pillars, is also in agreement with this argument. This is furthermore highlighted by the rule of mixture analysis, where the observed strength of the composite pillars is much lower than the predicted one. The post mortem examination of the deformed pillars also strengthens the claim where we see almost all the deformation is limited to the top of the M.O. pillars (Fig. 4.9 b) and distinct slip steps due to the lack of any dislocation multiplication mechanisms. On the contrary, the deformation in S.F. pillars is more uniform throughout the height of the pillar which can be argued to be due to dislocation generation mechanism at the interface.

It can further be argued that the interface controlling the deformation process in these directionally solidified eutectics is also responsible for the independence of hardness values with respect to the processing conditions of the DS eutectics as observed in this study. The strength of the matrix and the fibers increase with the increase in the solidification speed and decrease in the fiber diameter and fiber spacing while the interface area per unit volume is also increasing which would provide more sites for dislocation generation, counteracting the increase in the strength due to decreasing length scales of the fiber and the matrix confined between them.

It should be mentioned that there appears to be a contradiction between the strength values obtained by nanoindentation and the pillar compressions experiments, in terms of the trends with respect to solidification speeds. The strength values observed in the nanoindentation experiments is about 2-3 times of that observed in the pillar compression tests, what has been observed for many metallic materials.

The nanoindentation experiments show a non-dependence of hardness with respect to the solidification speed. While, pillar compression tests for M.F. pillars show an increase in strength with increasing solidification speed samples. It should be mentioned that this trend is rather weak and that the error bar for the 80 mm/h is sample is fairly large. A possible explanation of this trend maybe that the interaction between the dislocation sources during compression of higher solidification speed M.F. pillars as the spacing between the fibers decreases from 20 mm/h samples to 80 mm/h samples causing increase in stress levels required to deform the pillars. In contrast, there is no effect of the solidification speed on the hardness in nanoindentation. This finding may relate to the stress state, which is inhomogeneous and triaxial vs homogeneous and uniaxial in the micro compression test. As a result, the easier dislocation nucleation related to the higher interface density may counterbalance the more difficult dislocation propagation.

The deformation mechanism in the NiAl-Mo alloys, as discussed in the work of Kwon et al. (Kwon et al., 2015), appears to differ from what is observed in this study for NiAl-Cr alloys. They have observed that in NiAl-Mo directionally solidified eutectic alloys, the NiAl matrix deforms first which leads to the accumulation of <100> type dislocations at the interface which triggers the <111> dislocation in the Mo fibers. The reason for this difference can be attributed to two main factors; 1) a misfit dislocation network surrounding the Mo fibers was not observed in their study. 2) The Mo fibers are much stronger compared to the NiAl matrix. Therefore, it is obvious in the case of NiAl-Mo that NiAl matrix starts deforming first, while in case NiAl-Cr, the Cr fibers show comparable or lower strengths when compared to the NiAl matrix. In both cases, dislocation nucleation at the interface may support the plastic deformation of the eutectics leading eventually to an increase in toughness.

5 Summary and outlook

Directionally solidified NiAl-Cr eutectics, prepared at three different solidification speed 20 mm/h, 50 mm/h, and 80 mm/h, were mechanically characterized using different micromechanical techniques. A summary of the different experiments performed and results in this study is presented in this chapter along with a comprehensive summary of the interpretation of these results. A short outlook on future steps is also presented.

5.1 Summary and interpretation of results

The key observation from the different micromechanical tests conducted in this project can be summarized as:

1. The microstructural examination of the samples prepared at different solidification speed reveals a composite microstructure with NiAl forming the matrix and Cr long continuous fibers. All the fibers are oriented along the growth direction. TEM examination by Antje Krüger (RA V, IMD graduate school) reveal that both matrix and fiber have a [001] growth direction. The fiber diameter and spacing between the fibers is inversely related to the solidification speed. Both fiber diameter and spacing between the fibers decrease with increasing solidification speed.
2. Nanoindentation examination was carried out on the samples with varying solidification speed to evaluate Hardness and Young's modulus. The experiments reveal that the hardness and Young's modulus does not depend on the solidification speed of the directional solidification. This observation suggests that the deformation in these directionally solidified eutectics involving different phases is a complex process and individual phases and

the interface must be tested in order to understand the contribution of different phases in the deformation process.

3. The in-situ SEM tensile tests of the isolated individual Cr fibers revealed that the Cr fibers can be deformed plastically, at least in the given diameter range, and show high strength values up to 3 GPa. A size effect on the strength of the Cr fibers was observed, with the maximum strength of the Cr fibers increase with the decreasing fiber diameter. Furthermore, some of the long enough broken segments were re-gripped and tested again. The re-gripped segments show higher strength values than the pristine fibers of the same diameter.
4. Micro pillar compression tests were carried out on the three different kind of pillars, i.e. matrix only M.O. pillars, single fiber composite S.F. pillars, and size independent multiple fiber M.F. pillars on different solidification speed samples. The strength of the M.O. pillars and S.F. pillars increases with increasing solidification speed of the samples, due to decreasing fiber diameter and spacing between the fibers as evident from the microstructure examinations. The pillars show high strength values up to 4 GPa. The M.F. pillars prepared for all three samples show that the 0.2% yield strength is lowest of all the three types of pillars and increases weakly with increasing solidification speed.
   Comparing the strengths of single phase NiAl micro-pillars, having similar pillar diameter as S.F. pillars, reveals that for the same size the M.O. pillars are stronger than the S.F. pillars.
5. The post mortem examination of the deformed pillars by SEM reveals distinct slip steps in the M.O. pillars with almost all the deformation concentrated at the top of the pillar. In contrast, for the S.F. pillars, the deformation is more uniform throughout the height of the pillars. The cross cut across the deformed S.F. pillars further reveal that both the fiber and matrix is co-deformed and there was no sign of delamination or fracture at the interface. Examination of multiple fiber composite M.F. pillars reveal that all

the pillars failed by buckling related to a plastic buckling instability. The plastic instability occurs after yielding at around 15 % strain possibly due to misalignment between the indenter tip and sample surface or the large aspect ratio of the fibers.

6. TEM comparisons of undeformed and deformed S.F. pillars reveals that in the undeformed specimen the NiAl matrix contains small number of dislocations, Cr fibers are essentially dislocation free and the interface between the matrix and the fiber is surrounded by the network of <100> type interface dislocation arising due to semi-coherent interface between the two phases. On the other hand, the deformed pillars reveal much more dislocation activity, both inside the matrix and the fiber. Most of these dislocations were identified as <100> type, while dislocations with mixed Burgers vector were also observed.

The high strength values as well as the size effect in strength of these Cr fibers suggests that the Cr fibers have a very low initial dislocation density. This is also confirmed by the TEM examination where almost no dislocations were observed in the undeformed Cr fibers. The increase in the strength of the re-attached Cr fibers also supports this argument as the deformation in such fibers follows a weakest link approach.

The observed size effect in all the M.O. and S.F. has been attributed to dimensional constraints. The distinct slip steps in the deformed M.O. pillars suggest that M.O. pillars have a low initial dislocation density. In contrast, the more uniform deformation of S.F. pillars suggests stronger dislocation activity which is also confirmed by the TEM examination of the deformed S.F. pillars. The absence of any crack or delamination at the interface in the cross-cut across of the deformed S.F. pillars suggests that the interface between the fiber and the matrix is very strong and leads to the co-deformation of fiber and matrix. The lower strength of S.F. pillar, containing the interface, as compared to M.O. pillar, without the interface, of similar diameter suggests that the interface between the NiAl and Cr fibers plays an important role in the deformation of NiAl-Cr eutectic. This argument is further supported by the rule of mixture analysis, performed to estimate the strength of S.F. pillars based on the strength of Cr fiber and M.O. pillars. The analysis shows that the observed strength of the S.F. pillars is lower than it would be predicted by the rule of mixture.

All the observations from the different micromechanical tests conducted in this study indicate that the interface between the Cr fiber and NiAl matrix in directionally solidified NiAl-Cr eutectic, containing interface dislocation network, dominates the deformation mechanism of the eutectic alloys by generating dislocations when the two phases are co-deformed. This mechanism provides additional ductility to the brittle NiAl matrix and possibly results in increased toughness by delaying the crack nucleation or propagation.

5.2 Outlook

The observations and results from different micromechanical tests performed in this study indicates that the interface between the fiber and the matrix dominates the deformation mechanisms in NiAl-Cr directional solidified eutectic alloy by generating dislocations. This should provide additional toughness for the alloy. For a proof of principle, a study on the fracture behaviour of the alloy and the role of ductile Cr fibers and their orientation would be required. Being a first generation Ph.D. within the framework of IMD, the main focus of this study was to better understand the deformation processes in these alloys at room temperature. But NiAl-Cr being a candidate for high temperature application, its high temperature deformation behaviour must also be studied in detail. The following possible approaches are suggested to achieve the above mentioned aims:

1. To study the direct evidence of dislocation generation mechanism at the interface, a variant of micro-pillar compression test could be employed, where the tests are interrupted at a suitable indenter displacement such that the pillars just start to deform plastically, i.e. to deform the pillars partially. The lamellae from

these partially deformed pillars should then be examined by TEM to observe the dislocation evolution process.

2. To understand the effect of the ductile fibers on the fracture behaviour of the DS eutectic alloys, small scale fracture samples could be prepared and tested in-situ in order to observe the crack propagation process. As a starting point, one of the fracture samples based on the double cantilever beam (DCB) design of Liu et al. (S. Liu et al., 2013) were prepared by FIB milling to be tested in-situ (Fig. 5.1). High beam current was used to mill the outer shape of the sample. Special mounting blocks were prepared in which the sample can be mounted at an angle of 52°with respect to the horizontal axis and by manipulating the stage tilt the sample can be oriented parallel to the ion beam, thus allowing the middle section and crack was milled from the side to ensure a good surface finish. Lower beam currents were used to mill the middle section of and the crack. The middle section of the sample is ~4 µm X 4 µm in area. Crack depth is ~ 8 µm. The µ-DCB samples after preparation by FIB milling are compressed in-situ with a flat punch with diameter larger than 5 µm, such that both the shoulders of the samples are compressed simultaneously. The samples presented in figure 5.1 has the crack orientation parallel to the growth direction of the fiber. The µ-DCB sample can also be prepared in different orientations such that the orientation of fiber with respect to the crack can be varied in order to study the effect of fiber orientation on the fracture behaviour.
3. In addition to room temperature studies carried out in this study, high temperature deformation behaviour of the eutectic alloys could also be studied at small scale to understand the effect of fiber and their orientation with respect to the loading direction on the deformation and creep behaviour at elevated temperature.

Figure 5.1: µ-DCB fracture sample prepared by FIB milling, to be tested in-situ in order to study the crack propagation in DS eutectics with respect to the different fiber orientation. The sample prepared has the crack running parallel to the fiber orientation.

4. As one of the important goals of the graduate school IMD, effort must be directed towards developing different alloys prepared by adding one or more refractory metals to NiAl-Cr eutectic and the effect of these alloying additions on the overall behaviour of these alloys should also be evaluated.

References

Ansara, I., Dupin, N., Lukas, H. L., & Sundman, B. (1997). Thermodynamic assessment of the Al-Ni system. *Journal of Alloys and Compounds, 247*(1–2), 20–30. https://doi.org/10.1016/S0925-8388(96)02652-7

Anton, D. L., Shah, D. M., Duhl, D. N., & Giamei, A. F. (1989). Selecting high-temperature structural intermetallic compounds: The engineering approach. *JOM, 41*(9), 12–17. https://doi.org/10.1007/BF03220324

Arzt, E. (1998). Size effects in materials due to microstructural and dimensional constraints: A comparative review. *Acta Materialia, 46*(16), 5611–5626. https://doi.org/Doi 10.1016/S1359-6454(98)00231-6

Ashby, M. F. (1970). The deformation of plastically non-homogeneous materials. *Philosophical Magazine, 21*(170), 399–424. https://doi.org/10.1080/14786437008238426

Ball, A., & Smallman, R. . (1966a). The deformation properties and electron microscopy studies of the intermetallic compound NiAl. *Acta Metallurgica, 14*(10), 1349–1355. https://doi.org/10.1016/0001-6160(66)90251-3

Ball, A., & Smallman, R. . (1966b). The operative slip system and general plasticity of NiAl-II. *Acta Metallurgica, 14*(11), 1517–1526. https://doi.org/10.1016/0001-6160(66)90173-8

Bei, H., & George, E. P. (2005). Microstructures and mechanical properties of a directionally solidified NiAl-Mo eutectic alloy. *Acta Materialia, 53*(1), 69–77. https://doi.org/10.1016/j.actamat.2004.09.003

Bei, H., Pharr, G. M., & George, E. P. (2004). A review of directionally solidified intermetallic composites for high-temperature structural applications. *Journal of Materials Science*, *39*, 3975–3984. https://doi.org/10.1023/b:jmsc.0000031479.32138.84

Bei, H., Shim, S., George, E. P., Miller, M. K., Herbert, E. G., & Pharr, G. M. (2007). Compressive strengths of molybdenum alloy micro-pillars prepared using a new technique. *Scripta Materialia*, *57*(5), 397–400. https://doi.org/10.1016/j.scriptamat.2007.05.010

Bei, H., Shim, S., Pharr, G. M., & George, E. P. (2008). Effects of pre-strain on the compressive stress-strain response of Mo-alloy single-crystal micropillars. *Acta Materialia*, *56*(17), 4762–4770. https://doi.org/10.1016/j.actamat.2008.05.030

Biery, N., de Graef, M., & Pollock, T. M. (2003). A method for measuring microstructural-scale strains using a scanning electron microscope: Applications to γ-titanium aluminides. *Metallurgical and Materials Transactions A*, *34*(10), 2301–2313. https://doi.org/10.1007/s11661-003-0294-7

Birol, Y. (2011). A novel C-free Co-based alloy for high temperature tooling applications. *Materials Science and Engineering A*, *528*(3), 1117–1124. https://doi.org/10.1016/j.msea.2010.10.037

Bowman, R., Noebe, R., Raj, S., & Locci, I. (1992). Correlation of deformation mechanisms with the tensile and compressive behavior of NiAl and NiAl (Zr) intermetallic alloys. *Metallurgical Transactions A*, *23A*(May), 1493–1508. Retrieved from http://link.springer.com/article/10.1007/BF02647332

Bowman, R. R., & Noebe, R. D. (1992). Development of NiAl and NiAl-based composites for strucutral applications: A status report. *Superalloys 1992*, 341–350.

Brenner, S. S. (1956). Tensile strength of whiskers. *Journal of Applied Physics*, *27*(12), 1484–1491. https://doi.org/10.1063/1.1722294

Brenner, S. S. (1957). Plastic deformation of copper and silver whiskers. *Journal of Applied Physics*, *28*(9), 1023–1026. https://doi.org/10.1063/1.1722900

Brinckmann, S., Kim, J. Y., & Greer, J. R. (2008). Fundamental differences in mechanical behavior between two types of crystals at the nanoscale. *Physical Review Letters, 100*(15). https://doi.org/10.1103/PhysRevLett.100.155502

Chang, K. M., Darolia, R., & Lipsitt, H. A. (1992). Cleavage fracture in B2 aluminides. *Acta Metallurgica Et Materialia, 40*(10), 2727–2737. https://doi.org/10.1016/0956-7151(92)90343-D

Chen, W., Hines, J. R., & Wang, Y. (2004). Significant Improvement in Room Temperature Ductility of NiAl by Cr-Ce Complexes. *Advanced Engineering Materials, 6*(11), 876–879. https://doi.org/10.1002/adem.200400095

Chen, X. F., Johnson, D. R., Noebe, R. D., & Oliver, B. F. (1995). Deformation and fracture of a directionally solidified NiAl–28Cr–6Mo eutectic alloy. *Journal of Materials Research, 10*(5), 1159–1170. https://doi.org/10.1557/JMR.1995.1159

Chia, K.-H., Jung, K., & Conrad, H. (2005). Dislocation density model for the effect of grain size on the flow stress of a Ti–15.2 at.% Mo β-alloy at 4.2–650K. *Materials Science and Engineering: A, 409*(1–2), 32–38. https://doi.org/10.1016/j.msea.2005.03.117

Cline, E., Walter, J. L., Koch, E. F., & Osika, L. M. (1971). The variation of interface dislocation networks with lattice mismatch in eutectic alloys. *Acta Metallurgica, 19*, 405–414.

Cline, H. E., Walter, J. L., Lifshin, E., & Russell, R. R. (1971). Structures, faults, and the rod-plate transition in eutectics. *Metallurgical Transactions, 2*(1), 189–194. https://doi.org/10.1007/BF02662656

Darolia, R. (1991). NiAI Alloys for High. Temperature Structural Applications. *JOM, 43*(3), 44–48.

Deevi, S. C., & Sikka, V. K. (1996). Nickel and iron aluminides: An overview on properties, processing, and applications. *Intermetallics, 4*(5), 357–375. https://doi.org/10.1016/0966-9795(95)00056-9

Dey, G. K. (2003). Physical metallurgy of nickel aluminides. *Sadhana, 28*(1–2), 247–262. https://doi.org/10.1007/BF02717135

Dimiduk, D. M., Uchic, M. D., & Parthasarathy, T. A. (2005). Size-affected single-slip behavior of pure nickel microcrystals. *Acta Materialia, 53*(15), 4065–4077. https://doi.org/10.1016/j.actamat.2005.05.023

Duesbery, M. S., & Vitek, V. (1998). Plastic anisotropy in b.c.c. transition metals. *Acta Materialia, 46*(5), 1481–1492. https://doi.org/10.1016/S1359-6454(97)00367-4

Eberl, C., Gianola, D., & Thompson, R. (2006). Digital Image Correlation and Tracking. Retrieved from http://de.mathworks.com/matlabcentral/fileexchange/12413-digital-image-correlation-and-tracking

Fei, H., Abraham, A., Chawla, N., & Jiang, H. (2012). Evaluation of Micro-Pillar Compression Tests for Accurate Determination of Elastic-Plastic Constitutive Relations. *Journal of Applied Mechanics, 79*(6), 61011. https://doi.org/10.1115/1.4006767

Fischer-Cripps, A. C. (2004). *Nanoindentation*. New York, NY: Springer New York. https://doi.org/10.1007/978-1-4757-5943-3

Frank, F. C., & van der Merwe, J. H. (1949). One-Dimensional Dislocations. II. Misfitting Monolayers and Oriented Overgrowth. *Proceedings of the Royal Society A: Mathematical, Physical and Engineering Sciences, 198*(1053), 216–225. https://doi.org/10.1098/rspa.1949.0096

Frommeyer, G., & Rablbauer, R. (2008). High temperature materials based on the intermetallic compound NiAl reinforced by refractory metals for advanced energy conversion technologies. *Steel Research Int., 79*(7), 507–513. https://doi.org/10.2374/SRI08SP101-79-2008-507

Frommeyer, G., Rablbauer, R., & Schäfer, H. J. (2010). Elastic properties of B2-ordered NiAl and NiAl-X (Cr, Mo, W) alloys. *Intermetallics, 18*(3), 299–305. https://doi.org/10.1016/j.intermet.2009.07.026

Gianola, D. S., & Eberl, C. (2009). Micro- and nanoscale tensile testing of materials. *JOM, 61*(3), 24–35. https://doi.org/10.1007/s11837-009-0037-3

Gianola, D. S., Sedlmayr, a., Mnig, R., Volkert, C. a., Major, R. C., Cyrankowski, E., ... Kraft, O. (2011). In situ nanomechanical testing in focused ion beam and scanning electron microscopes. *Review of Scientific Instruments, 82*(6). https://doi.org/10.1063/1.3595423

Greer, J. R., & De Hosson, J. T. M. (2011). Plasticity in small-sized metallic systems: Intrinsic versus extrinsic size effect. *Progress in Materials Science, 56*(6), 654–724. https://doi.org/10.1016/j.pmatsci.2011.01.005

Greer, J. R., & Nix, W. D. (2006). Nanoscale gold pillars strengthened through dislocation starvation. *Physical Review B, 73*(24), 245410. https://doi.org/10.1103/PhysRevB.73.245410

Greer, J. R., Oliver, W. C., & Nix, W. D. (2005). Size dependence of mechanical properties of gold at the micron scale in the absence of strain gradients. *Acta Materialia, 53*(6), 1821–1830. https://doi.org/10.1016/j.actamat.2004.12.031

Greer, J. R., Weinberger, C. R., & Cai, W. (2008). Comparing the strength of f.c.c. and b.c.c. sub-micrometer pillars: Compression experiments and dislocation dynamics simulations. *Materials Science and Engineering A, 493*(1–2), 21–25. https://doi.org/10.1016/j.msea.2007.08.093

Gruber, P. A., Solenthaler, C., Arzt, E., & Spolenak, R. (2008). Strong single-crystalline Au films tested by a new synchrotron technique. *Acta Materialia, 56*(8), 1876–1889. https://doi.org/10.1016/j.actamat.2007.12.043

Hack, J. E., Brzeski, J. M., & Darolia, R. (1995). Fracture and deformation of NiAl single crystals. *Materials Science and Engineering: A, 192–193*, 268–276. https://doi.org/10.1016/0921-5093(94)03212-2

Haenschke, T., Gali, a, Heilmaier, M., Krüger, M., Bei, H., & George, E. P. (2010). Synthesis and characterization of lamellar and fibre-reinforced NiAl-Mo and NiAl-Cr. *Journal of Physics: Conference Series, 240*, 12063. https://doi.org/10.1088/1742-6596/240/1/012063

Hahn, K. H., & Vedula, K. (1989). Room temperature tensile ductility in polycrystalline B2 NiAl. *Scripta Metallurgica, 23*(1), 7–12. https://doi.org/10.1016/0036-9748(89)90083-5

Hall, E. O. (1951). The Deformation and Ageing of Mild Steel: III Discussion of Results. *Proceedings of the Physical Society. Section B, 64*(9), 747–753. https://doi.org/10.1088/0370-1301/64/9/303

Han, S. M., Bozorg-Grayeli, T., Groves, J. R., & Nix, W. D. (2010). Size effects on strength and plasticity of vanadium nanopillars. *Scripta Materialia, 63*(12), 1153–1156. https://doi.org/10.1016/j.scriptamat.2010.08.011

Hommel, M., Kraft, O., & Arzt, E. (2017). A new method to study cyclic deformation of thin films in tension and compression. https://doi.org/10.1557/JMR.1999.0317

Hutchinson, J. W. (2000). Plasticity at the micron scale. *Int. J. Solids Structures, 37*, 225–238.

Hütsch, J., & Lilleodden, E. T. (2014). The influence of focused-ion beam preparation technique on microcompression investigations: Lathe vs. annular milling. *Scripta Materialia, 77*, 49–51. https://doi.org/10.1016/j.scriptamat.2014.01.016

Jackson, K., & Hunt, J. (1966). Lamellar and rod eutectic growth. *AIME MET SOC TRANS, 236*(8), 1129. Retrieved from http://www.csa.com/partners/viewrecord.php?requester=gs&collection=TRD&recid=6611144475MD%5Cnhttp://scholar.google.com/scholar?hl=en&btnG=Search&q=intitle:Lamellar+and+rod+eutectic+growth#0

Johanns, K. E., Sedlmayr, A., Sudharshan Phani, P., Mönig, R., Kraft, O., George, E. P., & Pharr, G. M. (2012). In-situ tensile testing of single-crystal molybdenum-alloy fibers with various dislocation densities in a scanning electron microscope. *Journal of Materials Research, 27*(3), 508–520. https://doi.org/10.1557/jmr.2011.298

Johnson, D. R. (1994). Processing and Mechanical Properties of NiAI-Based In-Situ Composites.

Johnson, D. R., Chen, X. F., Oliver, B. F., Noebe, R. D., & Whittenberger, J. D. (1995a). Directional Solidification and Mechanical Properties of NiAl-NiAlTa Alloys, *3*, 141–152. https://doi.org/10.1016/0966-9795(95)92679-T

Johnson, D. R., Chen, X. F., Oliver, B. F., Noebe, R. D., & Whittenberger, J. D. (1995b). Processing and Mechanical Properties of In-situ Composites from the NiAl-Cr and the NiAl-(Cr,Mo) Eutectic Systems, *3*. https://doi.org/10.1016/0966-9795(95)92674-O

Kang, W., Rajagopalan, J., & Saif, M. T. A. (2010). In situ uniaxial mechanical testing of small scale materials - A review. *Nanosci. Nanotechnol. Lett.*, *2*(4), 282–287. https://doi.org/10.1166/nnl.2010.1107

Kiener, D., Grosinger, W., Dehm, G., & Pippan, R. (2008). A further step towards an understanding of size-dependent crystal plasticity: In situ tension experiments of miniaturized single-crystal copper samples. *Acta Materialia*, *56*(3), 580–592. https://doi.org/10.1016/j.actamat.2007.10.015

Kiener, D., Motz, C., & Dehm, G. (2009). Micro-compression testing: A critical discussion of experimental constraints. *Materials Science and Engineering A*, *505*(1–2), 79–87. https://doi.org/10.1016/j.msea.2009.01.005

Kiener, D., Motz, C., Schöberl, T., Jenko, M., & Dehm, G. (2006). Determination of Mechanical Properties of Copper at the Micron Scale. *Advanced Engineering Materials*, *8*(11), 1119–1125. https://doi.org/10.1002/adem.200600129

Kiener, D., Motz, C., Schobert, T., Jenko, M., & Dehm, G. (2006). Determination of mechanical properties of copper at the micron scale. *Advanced Engineering Materials*, *8*(11), 1119–1125. https://doi.org/10.1002/adem.200600129

Kim, J. Y., & Greer, J. R. (2009). Tensile and compressive behavior of gold and molybdenum single crystals at the nano-scale. *Acta Materialia*, *57*(17), 5245–5253. https://doi.org/10.1016/j.actamat.2009.07.027

Kim, J. Y., Jang, D., & Greer, J. R. (2012). Crystallographic orientation and size dependence of tension-compression asymmetry in molybdenum nano-pillars. *International Journal of Plasticity*, *28*(1), 46–52. https://doi.org/10.1016/j.ijplas.2011.05.015

Kraft, O., Gruber, P. a., Mönig, R., & Weygand, D. (2010). Plasticity in Confined Dimensions. *Annual Review of Materials Research*, *40*(1), 293–317. https://doi.org/10.1146/annurev-matsci-082908-145409

Kraft, O., Hommel, M., & Arzt, E. (2000). X-ray diffraction as a tool to study the mechanical behaviour of thin films. In *Materials Science and Engineering A* (Vol. 288, pp. 209–216). https://doi.org/10.1016/S0921-5093(00)00876-5

Kraft, O., & Volkert, C. A. (2001). Mechanical Testing of Thin Films and Small Structures. *Advanced Engineering Materials*, *3*(3), 99–110. https://doi.org/10.1002/1527-2648(200103)3:3<99::AID-ADEM99>3.0.CO;2-2

Kubaschewski, O. (1958). The heats of formation in the system aluminium + nickel + titanium. *Transactions of the Faraday Society*, *54*(0), 814. https://doi.org/10.1039/tf9585400814

Kumar, K. S., Mannan, S. K., & Viswanadham, R. K. (1992). Fracture toughness of NiAl and NiAl-based composites. *Acta Metallurgica et Materialia*, *40*(6), 1201–1222. https://doi.org/10.1016/0956-7151(92)90419-F

Kurz, W., & Fisher, D. J. (1986). *Fundamentals of solidification*. Trans Tech Publications Ltd, Trans Tech House, 4711, Aedermannsdorf, Switzerland, 1986.

Kwon, J., Bowers, M. L., Brandes, M. C., McCreary, V., Robertson, I. M., Phani, P. S., … Mills, M. J. (2015). Characterization of dislocation structures and deformation mechanisms in as-grown and deformed directionally solidified NiAl-Mo composites. *Acta Materialia*, *89*, 315–326. https://doi.org/10.1016/j.actamat.2015.01.059

Lahrman, D. F., Field, R. D., & Darolia, R. (1990). The Effect of Strain Rate on the Mechanical Properties of Single Crystal NiAl. *MRS Proceedings, 213*, 603. https://doi.org/10.1557/PROC-213-603

Lautenschlager, E. ., Hughes, T., & Brittain, J. . (1967). Slip in hard-sphere CsCl Lmodels. *Acta Metallurgica, 15*(8), 1347–1357. https://doi.org/10.1016/0001-6160(67)90009-0

Lee, T. C., Robertson, I. M., & Birnbaum, H. K. (1990). An In Situ transmission electron microscope deformation study of the slip transfer mechanisms in metals. *Metallurgical Transactions A, 21*(9), 2437–2447. https://doi.org/10.1007/BF02646988

Legros, M., Gianola, D. S., & Motz, C. (2010). Quantitative In Situ Mechanical Testing in Electron Microscopes. *MRS Bulletin, 35*(5), 354–360. https://doi.org/10.1557/mrs2010.567

Li, J. C. M. (1963). Petch Relation and Grain Boundary Sources. *Transactions of the Metallurgical Society of AIME, 227*, 239.

Liu, C. T., & Horton, J. a. (1995). Effect of refractory alloying additions on mechanical properties of near-stoichiometric NiAl. *Materials Science and Engineering: A, 192–193*, 170–178. https://doi.org/10.1016/0921-5093(94)03232-7

Liu, S., Wheeler, J. M., Howie, P. R., Zeng, X. T., Michler, J., & Clegg, W. J. (2013). Measuring the fracture resistance of hard coatings. *Applied Physics Letters, 102*(17), 171907. https://doi.org/10.1063/1.4803928

Meetham, G. W. (1991). High-temperature materials—a general review. *Journal of Materials Science, 26*(1991), 853–860. https://doi.org/10.1007/BF00576759

Miracle, D. B. (1993). The physical and mechanical properties of NiAl. *Acta Metall. Mater., 41*(3), 649–684.

Miracle, D. B., & Darolia, R. (2000). NiAl and its Alloys. In *Structural Applications of Intermetallic Compounds* (pp. 55–74). Retrieved from http://www.ewp.rpi.edu/hartford/users/papers/engr/ernesto/morens/EP/References/NiAl.pdf

Misra, A., & Gibala, R. (2000). Plasticity in multiphase intermetallics. *Intermetallics*, *8*(9–11), 1025–1034. https://doi.org/10.1016/S0966-9795(00)00079-0

Misra, A., Wu, Z. L., Kush, M. T., & Gibala, R. (1998). Deformation and fracture behaviour of directionally solidified NiAl-Mo and NiAl-Mo(Re) eutectic composites. *Philosophical Magazine A*, *78*(3), 533–550. https://doi.org/10.1080/014186198253363

Mompiou, F., Legros, M., Sedlmayr, A., Gianola, D. S., Caillard, D., & Kraft, O. (2012). Source-based strengthening of sub-micrometer Al fibers. *Acta Materialia*, *60*(3), 977–983. https://doi.org/10.1016/j.actamat.2011.11.005

Mori, H., Fujita, H., Tendo, M., & Fujita, M. (1984). Amorphous transition in intermetallic compounds induced by electron irradiation. *Scripta Metallurgica*, *18*(8), 783–788. https://doi.org/10.1016/0036-9748(84)90394-6

Moser, B., Wasmer, K., Barbieri, L., & Michler, J. (2007). Strength and fracture of Si micropillars: A new scanning electron microscopy-based micro-compression test. *Journal of Materials Research*, *22*(4), 1004–1011. https://doi.org/10.1557/jmr.2007.0140

Noebe, R. D., Bowman, R. R., & Nathal, M. V. (1993). Physical and mechanical properties of the B2 compound NiAI. *International Materials Reviews*, *38*(4).

Noebe, R. D., Misra, a., & Gibala, R. (1991). Plastic Flow and Fracture of B2 NiAl-based Intermetallic Alloys Containing a Ductile Second Phase. *ISIJ International*, *31*(10), 1172–1185. https://doi.org/10.2355/isijinternational.31.1172

Oliver, W. C., & Pharr, G. M. (1992). An improved technique for determining hardness and elastic modulus using load and displacement sensing indentation experiments. *Journal of Materials Research*. https://doi.org/10.1557/JMR.1992.1564

Oliver, W. C., & Pharr, G. M. (2004). Measurement of hardness and elastic modulus by instrumented indentation: Advances in understanding and refinements to methodology. *Journal of Materials Research, 19*(1), 3–20. https://doi.org/10.1557/jmr.2004.19.1.3

Palacio, M. L. B., & Bhushan, B. (2013). Depth-sensing indentation of nanomaterials and nanostructures. *Materials Characterization, 78*, 1–20. https://doi.org/10.1016/j.matchar.2013.01.009

Pank, D. R., Koss, D. a., & Nathal, M. V. (1990). Microstructure and mechanical properties of multiphase NiAl-based alloys. *J. Mater. Res., 5*(5), 942–949. https://doi.org/10.1557/JMR.1990.0942

Pascoe, R. T., & Newwy, C. W. A. (1968). Deformation Modes of the Intermediate Phase NiAl. *Physica Status Solidi (A) Applications and Materials Science, 29*, 357.

Perepezko, J. H. (2009). The hotter the engine, the better. *Science, 326*(5956), 1068–1069. https://doi.org/10.1126/science.1179327

Petch, N. J. (1953). The Cleavage Strength of Polycrystals. *J. Iron Steel Inst, 174*, 25–28. https://doi.org/10.1007/BF01972547

Phani, P. S., Johanns, K. E., Duscher, G., Gali, a., George, E. P., & Pharr, G. M. (2011). Scanning transmission electron microscopc observations of defects in as-grown and pre-strained Mo alloy fibers. *Acta Materialia, 59*(5), 2172–2179. https://doi.org/10.1016/j.actamat.2010.12.018

Pollock, T. M., & Tin, S. (2006). Nickel-Based Superalloys for Advanced Turbine Engines: Chemistry, Microstructure and Properties. *Journal of Propulsion and Power, 22*(2), 361–374. https://doi.org/10.2514/1.18239

Rablbauer, R., Fischer, R., & Frommeyer, G. (2004). Mechanical properties of NiAl – Cr alloys in relation to microstructure and atomic defects. *Z. Metallkd., 95*(6), 525–534. Retrieved from http://www.ijmr.de/MK017973

Rachinger, W. ., & Cottrell, A. . (1956). Slip in crystals of the caesium chloride type. *Acta Metallurgica, 4*(2), 109–113. https://doi.org/10.1016/0001-6160(56)90128-6

Reed, R. C. (2006). *The Superalloys*. Cambridge: Cambridge University Press. https://doi.org/10.1017/CBO9780511541285

Reuss, S., & Vehoff, H. (1990). Temperature dependence of the fracture toughness of single phase and two phase intermetallics. *Scripta Metallurgica et Materialia, 24*(6), 1021–1026. https://doi.org/10.1016/0956-716X(90)90293-P

Richter, G., Hillerich, K., Gianola, D. S., Mönig, R., Kraft, O., & Volkert, C. A. (2009). Ultrahigh Strength Single Crystalline Nanowhiskers Grown by Physical Vapor Deposition. *Nano Letters, 9*(8), 3048–3052. https://doi.org/10.1021/nl9015107

Rusović, N., & Warlimont, H. (1977). The elastic behaviour of β2-NiAl alloys. *Physica Status Solidi (a), 44*(2), 609–619. https://doi.org/10.1002/pssa.2210440225

Rusović, N., & Warlimont, H. (1979). Young's modulus of β2-NiAl alloys. *Physica Status Solidi (a), 53*(1), 283–288. https://doi.org/10.1002/pssa.2210530132

Rzepiejewska-Malyska, K. A., Buerki, G., Michler, J., Major, R. C., Cyrankowski, E., Asif, S. A. S., & Warren, O. L. (2008). In situ mechanical observations during nanoindentation inside a high-resolution scanning electron microscope. *Journal of Materials Research, 23*(7), 1973–1979. https://doi.org/10.1557/JMR.2008.0240

Schneider, A. S., Clark, B. G., Frick, C. P., Gruber, P. a., & Arzt, E. (2009). Effect of orientation and loading rate on compression behavior of small-scale Mo pillars. *Materials Science and Engineering A, 508*(1–2), 241–246. https://doi.org/10.1016/j.msea.2009.01.011

Schneider, A. S., Kaufmann, D., Clark, B. G., Frick, C. P., Gruber, P. A., Mönig, R., Arzt, E. (2009). Correlation between critical temperature and strength of small-scale bcc pillars. *Physical Review Letters, 103*(10). Retrieved from https://journals.aps.org/prl/pdf/10.1103/PhysRevLett.103.105501

Schuh, C. A. (2006). Nanoindentation studies of materials. *Materials Today, 9*(5), 32–40. https://doi.org/10.1016/S1369-7021(06)71495-X

Schwaiger, R., Weber, M., Moser, B., Gumbsch, P., & Kraft, O. (2011). Mechanical assessment of ultrafine-grained nickel by microcompression experiment and finite element simulation. *Journal of Materials Research, 27*(1), 266–277. https://doi.org/10.1557/jmr.2011.248

Sedlmayr, A. (2012). *Experimental Investigations of Deformation Pathways in Nanowires*. Karlsruher Institut für Technologie (KIT).

Sedlmayr, A., Bitzek, E., Gianola, D. S., Richter, G., M??nig, R., & Kraft, O. (2012). Existence of two twinning-mediated plastic deformation modes in Au nanowhiskers. *Acta Materialia, 60*(9), 3985–3993. https://doi.org/10.1016/j.actamat.2012.03.018

Senger, J., Weygand, D., Motz, C., Gumbsch, P., & Kraft, O. (2011). Aspect ratio and stochastic effects in the plasticity of uniformly loaded micrometer-sized specimens. *Acta Materialia, 59*(8), 2937–2947. https://doi.org/10.1016/j.actamat.2011.01.034

Shan, Z. W., Mishra, R. K., Syed Asif, S. A., Warren, O. L., & Minor, A. M. (2008). Mechanical annealing and source-limited deformation in submicrometre-diameter Ni crystals. *Nature Materials, 7*(2), 115–119. https://doi.org/10.1038/nmat2085

Shen, Y.-L., Suresh, S., He, M. Y., Bagchi, a., Kienzle, O., Rühle, M., & Evans, a. G. (1998). Stress evolution in passivated thin films of Cu on silica substrates. *Journal of Materials Research, 13*(7), 1928–1937. https://doi.org/10.1557/JMR.1998.0272

Sutton, M. A., Li, N., Garcia, D., Cornille, N., Orteu, J. J., McNeill, S. R., ... Reynolds, A. P. (2007). Scanning Electron Microscopy for Quantitative Small and Large Deformation Measurements Part II: Experimental Validation for Magnifications from 200 to 10,000. *Experimental Mechanics, 47*(6), 789–804. https://doi.org/10.1007/s11340-007-9041-0

Sutton, M. A., Li, N., Joy, D. C., Reynolds, A. P., & Li, X. (2007). Scanning Electron Microscopy for Quantitative Small and Large Deformation Measurements Part I: SEM Imaging at Magnifications from 200 to 10,000. *Experimental Mechanics, 47*(6), 775–787. https://doi.org/10.1007/s11340-007-9042-z

Taylor, A., & Doyle, N. J. (1972). Further Studies on the Nickel-Aluminum System. I. The ß-NiAl and δ-Ni2Al3 Phase Fields. *Journal of Applied Crystallography, 5*, 201–209. https://doi.org/10.1107/S0021889872009203

Taylor, G. I. (1934). The Mechanism of Plastic Deformation of Crystals. Part I. Theoretical. *Proceedings of the Royal Society A: Mathematical, Physical and Engineering Sciences, 145*(855), 362–387. https://doi.org/10.1098/rspa.1934.0106

Uchic, M. D., & Dimiduk, D. M. (2005). A methodology to investigate size scale effects in crystalline plasticity using uniaxial compression testing. *Materials Science and Engineering A, 400–401*(1–2 SUPPL.), 268–278. https://doi.org/10.1016/j.msea.2005.03.082

Uchic, M. D., Dimiduk, D. M., Florando, J. N., & Nix, W. D. (2004). Sample dimensions influence strength and crystal plasticity. *Science (New York, N.Y.), 305*(5686), 986–989. https://doi.org/10.1126/science.1098993

Uchic, M. D., Shade, P. a., & Dimiduk, D. M. (2009a). Micro-compression testing of fcc metals: A selected overview of experiments and simulations. *JOM, 61*(3), 36–41. https://doi.org/10.1007/s11837-009-0038-2

Uchic, M. D., Shade, P. a., & Dimiduk, D. M. (2009b). Plasticity of Micrometer-Scale Single Crystals in Compression. *Annual Review of Materials Research*, *39*(1), 361–386. https://doi.org/10.1146/annurev-matsci-082908-145422

Villars, P. (Pierre), Calvert, L. D. (Lauriston D.., & Pearson, W. B. (William B. (1985). *Pearson's handbook of crystallographic data for intermetallic phases*. American Society for Metals.

Walter, J. L., & Cline, H. E. (1970). The effect of solidification rate on structure and high-temperature strength of the eutectic NiAl-Cr. *Metallurgical and Materials Transactions*, *1*(5), 1221–1229. https://doi.org/10.1007/BF02900234

Walter, J. L., & Cline, H. E. (1973). Stability of the directionally solidified eutectics NiAl-Cr and NiAl-Mo. *Metallurgical Transactions*, *4*(1), 33–38. https://doi.org/10.1007/BF02649602

Wang, J., Qian, J., Zhang, X., & Wang, Y. (2011). Research status and progress of NiAl based alloys as high temperature structural materials. *Rare Metals*, *30*(SUPPL.1), 422–426. https://doi.org/10.1007/s12598-011-0317-2

Weinberger, C. R., & Cai, W. (2008). Surface-controlled dislocation multiplication in metal micropillars. *Proceedings of the National Academy of Sciences of the United States of America*, *105*(38), 14304–14307. https://doi.org/10.1073/pnas.0806118105

Whittenberger, J. D., Raj, S. V., Locci, I. E., & Salem, J. a. (1999). Effect of growth rate on elevated temperature plastic flow androom temperature fracture toughness of directionally solidified NiAl-31Cr-3Mo. *Intermetallics*, *7*(10), 1159–1168. https://doi.org/10.1016/S0966-9795(99)00023-0

Yang, J.-M. (1997a). Microstructure and mechanical behavior if in-situ directional solidified NiAl/Cr(Mo) eutectic composite, *45*(1), 295–305.

Yang, J.-M. (1997b). The mechanical behavior of in-situ NiAl-refractory metal composites. *JOM*. https://doi.org/10.1007/BF02914401

Yoo, M. H., Takasugi, T., Hanada, S., & Izumi, O. (1990). Slip Modes in B2-Type Intermetallic Alloys. *Materials Transactions, JIM*, *31*(6), 435–442. https://doi.org/10.2320/matertrans1989.31.435

Zhang, D., Breguet, J. M., Clavel, R., Sivakov, V., Christiansen, S., & Michler, J. (2010). In situ electron microscopy mechanical testing of silicon nanowires using electrostatically actuated tensile stages. *Journal of Microelectromechanical Systems*, *19*(3), 663–674. https://doi.org/10.1109/JMEMS.2010.2044746

Zhu, T. T., Bushby, A. J., & Dunstan, D. J. (2008). Materials mechanical size effects: a review. *Materials Technology*, *23*(4), 193–209. https://doi.org/10.1179/175355508X376843

Zhu, Y., & Espinosa, H. D. (2005). An electromechanical material testing system for in situ electron microscopy and applications. *Proceedings of the National Academy of Sciences of the United States of America*, *102*(41), 14503–8. https://doi.org/10.1073/pnas.0506544102

Schriftenreihe des Instituts für Angewandte Materialien

ISSN 2192-9963

Band 1 Prachai Norajitra
Divertor Development for a Future Fusion Power Plant. 2011
ISBN 978-3-86644-738-7

Band 2 Jürgen Prokop
Entwicklung von Spritzgießsonderverfahren zur Herstellung von Mikrobauteilen durch galvanische Replikation. 2011
ISBN 978-3-86644-755-4

Band 3 Theo Fett
New contributions to R-curves and bridging stresses – Applications of weight functions. 2012
ISBN 978-3-86644-836-0

Band 4 Jérôme Acker
Einfluss des Alkali/Niob-Verhältnisses und der Kupferdotierung auf das Sinterverhalten, die Strukturbildung und die Mikrostruktur von bleifreier Piezokeramik ($K_{0,5}Na_{0,5}$)NbO_3. 2012
ISBN 978-3-86644-867-4

Band 5 Holger Schwaab
Nichtlineare Modellierung von Ferroelektrika unter Berücksichtigung der elektrischen Leitfähigkeit. 2012
ISBN 978-3-86644-869-8

Band 6 Christian Dethloff
Modeling of Helium Bubble Nucleation and Growth in Neutron Irradiated RAFM Steels. 2012
ISBN 978-3-86644-901-5

Band 7 Jens Reiser
Duktilisierung von Wolfram. Synthese, Analyse und Charakterisierung von Wolframlaminaten aus Wolframfolie. 2012
ISBN 978-3-86644-902-2

Band 8 Andreas Sedlmayr
Experimental Investigations of Deformation Pathways in Nanowires. 2012
ISBN 978-3-86644-905-3

Die Bände sind unter www.ksp.kit.edu als PDF frei verfügbar oder als Druckausgabe bestellbar.

Band 9 Matthias Friedrich Funk
Microstructural stability of nanostructured fcc metals during cyclic deformation and fatigue. 2012
ISBN 978-3-86644-918-3

Band 10 Maximilian Schwenk
Entwicklung und Validierung eines numerischen Simulationsmodells zur Beschreibung der induktiven Ein- und Zweifrequenzrandschichthärtung am Beispiel von vergütetem 42CrMo4. 2012
ISBN 978-3-86644-929-9

Band 11 Matthias Merzkirch
Verformungs- und Schädigungsverhalten der verbundstranggepressten, federstahldrahtverstärkten Aluminiumlegierung EN AW-6082. 2012
ISBN 978-3-86644-933-6

Band 12 Thilo Hammers
Wärmebehandlung und Recken von verbundstranggepressten Luftfahrtprofilen. 2013
ISBN 978-3-86644-947-3

Band 13 Jochen Lohmiller
Investigation of deformation mechanisms in nanocrystalline metals and alloys by in situ synchrotron X-ray diffraction. 2013
ISBN 978-3-86644-962-6

Band 14 Simone Schreijäg
Microstructure and Mechanical Behavior of Deep Drawing DC04 Steel at Different Length Scales. 2013
ISBN 978-3-86644-967-1

Band 15 Zhiming Chen
Modelling the plastic deformation of iron. 2013
ISBN 978-3-86644-968-8

Band 16 Abdullah Fatih Çetinel
Oberflächendefektausheilung und Festigkeitssteigerung von niederdruckspritzgegossenen Mikrobiegebalken aus Zirkoniumdioxid. 2013
ISBN 978-3-86644-976-3

Band 17 Thomas Weber
Entwicklung und Optimierung von gradierten Wolfram/EUROFER97-Verbindungen für Divertorkomponenten. 2013
ISBN 978-3-86644-993-0

Die Bände sind unter www.ksp.kit.edu als PDF frei verfügbar oder als Druckausgabe bestellbar.

Band 18 Melanie Senn
Optimale Prozessführung mit merkmalsbasierter Zustandsverfolgung. 2013
ISBN 978-3-7315-0004-9

Band 19 Christian Mennerich
Phase-field modeling of multi-domain evolution in ferromagnetic shape memory alloys and of polycrystalline thin film growth. 2013
ISBN 978-3-7315-0009-4

Band 20 Spyridon Korres
On-Line Topographic Measurements of Lubricated Metallic Sliding Surfaces. 2013
ISBN 978-3-7315-0017-9

Band 21 Abhik Narayan Choudhury
Quantitative phase-field model for phase transformations in multi-component alloys. 2013
ISBN 978-3-7315-0020-9

Band 22 Oliver Ulrich
Isothermes und thermisch-mechanisches Ermüdungsverhalten von Verbundwerkstoffen mit Durchdringungsgefüge (Preform-MMCs). 2013
ISBN 978-3-7315-0024-7

Band 23 Sofie Burger
High Cycle Fatigue of Al and Cu Thin Films by a Novel High-Throughput Method. 2013
ISBN 978-3-7315-0025-4

Band 24 Michael Teutsch
Entwicklung von elektrochemisch abgeschiedenem LIGA-Ni-Al für Hochtemperatur-MEMS-Anwendungen. 2013
ISBN 978-3-7315-0026-1

Band 25 Wolfgang Rheinheimer
Zur Grenzflächenanisotropie von $SrTiO_3$. 2013
ISBN 978-3-7315-0027-8

Band 26 Ying Chen
Deformation Behavior of Thin Metallic Wires under Tensile and Torsional Loadings. 2013
ISBN 978-3-7315-0049-0

Band 27 Sascha Haller
Gestaltfindung: Untersuchungen zur Kraftkegelmethode. 2013
ISBN 978-3-7315-0050-6

Band 28 Nicht erschienen

Band 29 Gunnar Picht
Einfluss der Korngröße auf ferroelektrische Eigenschaften dotierter $Pb(Zr_{1-x}Ti_x)O_3$ Materialien. 2013
ISBN 978-3-7315-0106-0

Band 30 Esther Held
Eigenspannungsanalyse an Schichtverbunden mittels inkrementeller Bohrlochmethode. 2013
ISBN 978-3-7315-0127-5

Band 31 Pei He
On the structure-property correlation and the evolution of Nanofeatures in 12-13.5% Cr oxide dispersion strengthened ferritic steels. 2014
ISBN 978-3-7315-0141-1

Band 32 Jan Hoffmann
Ferritische ODS-Stähle – Herstellung, Umformung und Strukturanalyse. 2014
ISBN 978-3-7315-0157-2

Band 33 Wiebke Sittel
Entwicklung und Optimierung des Diffusionsschweißens von ODS Legierungen. 2014
ISBN 978-3-7315-0182-4

Band 34 Osama Khalil
Isothermes Kurzzeitermüdungsverhalten der hoch-warmfesten Aluminium-Knetlegierung 2618A (AlCu2Mg1,5Ni). 2014
ISBN 978-3-7315-0208-1

Band 35 Magalie Huttin
Phase-field modeling of the influence of mechanical stresses on charging and discharging processes in lithium ion batteries. 2014
ISBN 978-3-7315-0213-5

Band 36 Christoph Hage
Grundlegende Aspekte des 2K-Metallpulverspritzgießens. 2014
ISBN 978-3-7315-0217-3

Band 37 Bartłomiej Albiński
Instrumentierte Eindringprüfung bei Hochtemperatur für die Charakterisierung bestrahlter Materialien. 2014
ISBN 978-3-7315-0221-0

Band 38 Tim Feser
Untersuchungen zum Einlaufverhalten binärer alpha-Messinglegierungen unter Ölschmierung in Abhängigkeit des Zinkgehaltes. 2014
ISBN 978-3-7315-0224-1

Band 39 Jörg Ettrich
Fluid Flow and Heat Transfer in Cellular Solids. 2014
ISBN 978-3-7315-0241-8

Band 40 Melanie Syha
Microstructure evolution in strontium titanate Investigated by means of grain growth simulations and x-ray diffraction contrast tomography experiments. 2014
ISBN 978-3-7315-0242-5

Band 41 Thomas Haas
Mechanische Zuverlässigkeit von gedruckten und gasförmig abgeschiedenen Schichten auf flexiblem Substrat. 2014
ISBN 978-3-7315-0250-0

Band 42 Aron Kneer
Numerische Untersuchung des Wärmeübertragungsverhaltens in unterschiedlichen porösen Medien. 2014
ISBN 978-3-7315-0252-4

Band 43 Manuel Feuchter
Investigations on Joule heating applications by multiphysical continuum simulations in nanoscale systems. 2014
ISBN 978-3-7315-0261-6

Band 44 Alexander Vondrous
Grain growth behavior and efficient large scale simulations of recrystallization with the phase-field method. 2014
ISBN 978-3-7315-0280-7

Band 45 Tobias Kennerknecht
Fatigue of Micro Molded Materials – Aluminum Bronze and Yttria Stabilized Zirconia. 2014
ISBN 978-3-7315-0293-7

Band 46 Christopher Scherr
Elektrochemisches Verhalten von Lithium-Schwefel-Zellen mit unterschiedlicher Kathodenstruktur. 2015
ISBN 978-3-7315-0296-8

Band 47 Konstantin Frölich
Der Decal-Prozess zur Herstellung katalysatorbeschichteter Membranen für PEM-Brennstoffzellen. 2015
ISBN 978-3-7315-0334-7

Band 48 Benedikt Haspel
Werkstoffanalytische Betrachtung der Eigenschaften von mittels neuartiger RTM-Fertigungsprozesse hergestellten glasfaserverstärkten Polymerverbunden. 2015
ISBN 978-3-7315-0337-8

Band 49 Marco Berghoff
Skalenübergreifende Modellierung und Optimierung vom atomistischen kristallinen Phasenfeldmodell bis zur mesoskopischen Phasenfeldmethode. 2015
ISBN 978-3-7315-0416-0

Band 50 Michael Selzer
Mechanische und Strömungsmechanische Topologieoptimierung mit der Phasenfeldmethode. 2016
ISBN 978-3-7315-0431-3

Band 51 Michael Mahler
Entwicklung einer Auswertemethode für bruchmechanische Versuche an kleinen Proben auf der Basis eines Kohäsivzonenmodells. 2016
ISBN 978-3-7315-0441-2

Band 52 Christoph Bohnert
Numerische Untersuchung des Verformungs- und Bruchverhaltens von einkristallinem Wolfram auf mikroskopischer Ebene. 2016
ISBN 978-3-7315-0444-3

Band 53 Stefan Guth
Schädigung und Lebensdauer von Nickelbasislegierungen unter thermisch-mechanischer Ermüdungsbeanspruchung bei verschiedenen Phasenlagen. 2016
ISBN 978-3-7315-0445-0

Band 54 Markus Klinsmann
The Effects of Internal Stress and Lithium Transport on Fracture in Storage Materials in Lithium-Ion Batteries. 2016
ISBN 978-3-7315-0455-9

Band 55 Thomas Straub
Experimental Investigation of Crack Initiation in Face-Centered Cubic Materials in the High and Very High Cycle Fatigue Regime. 2016
ISBN 978-3-7315-0471-9

Band 56 Maren Lepple
Kupfer- und Eisenoxide als Konversions-Elektrodenmaterialien für Lithium-Ionen-Batterien: Thermodynamische und Elektrochemische Untersuchungen. 2016
ISBN 978-3-7315-0482-5

Band 57 Stefan Andreas Slaby
Charakterisierung und Bewertung der Zug- und Ermüdungseigenschaften von Mikrobauteilen aus 17-4PH Edelstahl. Ein Vergleich von mikropulverspritzgegossenem und konventionell hergestelltem Material. 2017
ISBN 978-3-7315-0484-9

Band 58 Kumar Ankit
Phase-field modeling of microstructural pattern formation in alloys and geological veins. 2016
ISBN 978-3-7315-0491-7

Band 59 Kuo Zhang
Characterization and Modeling of the Ratcheting Behavior of the Ferritic-Martensitic Steel P91. 2017
ISBN 978-3-7315-0503-7

Band 60 Jens Bauer
Mechanische Metamaterialien – Hohe Festigkeit und niedrige Dichte durch Mikroarchitektur und Materialdesign. 2017
ISBN 978-3-7315-0506-8

Band 61 Fabian Lemke
Untersuchung des Sinterverhaltens von $SrTiO_3$ unter Berücksichtigung der Defektchemie. 2016
ISBN 978-3-7315-0510-5

Band 62 Johannes Kümmel
Detaillierte Analyse der Aufbauschneidenbildung bei der Trockenzerspanung von Stahl C45E mit Berücksichtigung des Werkzeugverschleißes. 2016
ISBN 978-3-7315-0518-1

Band 63 László Hagymási
Modellierung der Stoffübertragung beim Niederdruck-carbonitrieren mit Ammoniak und Acetylen. 2016
ISBN 978-3-7315-0568-6

Band 64 Reza Eslami
A novel micro-mechanical model for prediction of multiaxial high cycle fatigue at small scales. 2017
ISBN 978-3-7315-0583-9

Band 65 Sebastian Schulz
Phase-field simulations of multi-component solidification and coarsening based on thermodynamic datasets. 2017
ISBN 978-3-7315-0618-8

Band 66 Markus Stricker
Die Übertragung von mikrostrukturellen Eigenschaften aus der diskreten Versetzungsdynamik in Kontinuumsbeschreibungen. 2017
ISBN 978-3-7315-0658-4

Band 67 Luis Straßberger
Untersuchung und Modellierung des viskoplastischen Verformungsverhaltens oxidpartikelverstärkter Stähle. 2017
ISBN 978-3-7315-0674-4

Band 68 Mark Wobrock
Microplasticity of idealized single crystalline Ag cantilevers characterized with methods of high resolution. 2017
ISBN 978-3-7315-0682-9

Band 69 Amritesh Kumar
Micromechanical study on the deformation behaviour of directionally solidified NiAl-Cr eutectic composites. 2017
ISBN 978-3-7315-0694-2

Die Bände sind unter www.ksp.kit.edu als PDF frei verfügbar oder als Druckausgabe bestellbar.